Survival

GLOBAL POLITICS AND STRATEGY
Volume 51 Number 1 | February–March 2009

The International Institute for Strategic Studies
Arundel House | 13–15 Arundel Street | Temple Place | London | WC2R 3DX | UK
Tel +44 (0)20 7379 7676 Fax +44 (0)20 7836 3108 E-mail survival@iiss.org Web www.iiss.org
Survival Online www.informaworld.com/survival

Aims and Scope *Survival* is one of the world's leading forums for analysis and debate of international and strategic affairs. Shaped by its editors to be both timely and forward thinking, the journal encourages writers to challenge conventional wisdom and bring fresh, often controversial, perspectives to bear on the strategic issues of the moment. With a diverse range of authors, *Survival* aims to be scholarly in depth while vivid, well written and policy relevant in approach. Through commentary, analytical articles, case studies, forums, review essays, reviews and letters to the editor, the journal promotes lively, critical debate on issues of international politics and strategy.

Published for the IISS by Routledge Journals,
2 Park Square, Milton Park, Abingdon, Oxon OX14 4RN
52 Vanderbilt Avenue, New York, NY 10017

Routledge is an imprint of the Taylor & Francis Group, an informa business

ISBN 13: 978-0-415-49838-8 (pbk)
ISBN 13: 978-1-138-45884-0 (hbk)

Survival
GLOBAL POLITICS AND STRATEGY
Volume 51 Number 1 | February–March 2009

Contents

Cover AP Photo/David Guttenfelder

On the cover
US Marines from the 24th Marine Expeditionary Unit try to take shelter from a sandstorm at forward operating base Dwyer in the Helmand province of southern Afghanistan, 7 May 2008.

New on the web
Visit http://www.iiss.org/publications/survival for Brief Notices on new books on the Middle East; Politics and International Relations; Arms, Arms Control and Technology; and the Asia-Pacific.

IISS Voices
A new blog from the editors of *Survival* and other IISS staff. Http://www.iiss.org/whats-new/iiss-voices.

Plus

Review Essay

Closing Argument

About the IISS The IISS, a registered charity with offices in London, Washington and Singapore, is the world's leading authority on political–military conflict. It is the primary independent source of accurate, objective information on international strategic issues. Publications include *The Military Balance*, an annual reference work on each nation's defence capabilities; *Strategic Survey*, an annual review of world affairs; *Survival*, a bi-monthly journal on international affairs; *Strategic Comments*, a monthly analysis of topical issues in international affairs; and the *Adelphi Paper* monograph series on policy-relevant strategic issues.

SUBMISSIONS

To submit an article, authors are advised to follow these guidelines:

- *Survival* articles are around 4,000–10,000 words long including endnotes. A word count should be included with a draft. Length is a consideration in the review process and shorter articles have an advantage.
- All text, including endnotes, should be double-spaced with wide margins.
- Any tables or artwork should be supplied in separate files, ideally not embedded in the document or linked to text around it.
- All *Survival* articles are expected to include endnote references. These should be complete and include first and last names of authors, titles of articles (even from newspapers), place of publication, publisher, exact publication dates, volume and issue number (if from a journal) and page numbers. Web sources should include complete URLs and DOIs if available.
- A summary of up to 150 words should be included with the article. The summary should state the main argument clearly and concisely, not simply say what the article is about.
- A short author's biography of one or two lines should also be included. This information will appear at the foot of the first page of the article.

Submissions should be made by email, in Microsoft Word format, to the Managing Editor, Jeffrey Mazo, survival@iiss.org. Alternatively, hard copies may be sent to *Survival*, IISS, Arundel House, 13–15 Arundel Street, Temple Place, London WC2R 3DX, UK. Please direct any queries to Jeffrey Mazo.

The editorial review process can take up to three months. *Survival*'s acceptance rate for unsolicited manuscripts is less than 20%. *Survival* does not normally provide referees' comments in the event of rejection. Authors are permitted to submit simultaneously elsewhere so long as this is consistent with the policy of the other publication and the Editors of *Survival* are informed of the dual submission.

Readers are encouraged to comment on articles from the previous issue. Letters should be concise, no longer than 750 words and relate directly to the argument or points made in the original article. A response from the author of the article will normally be solicited.

ADVERTISING AND PERMISSIONS

For advertising rates and schedules

USA/Canada: The Advertising Manager, Taylor & Francis Inc., 325 Chestnut Street, 8th Floor, Philadelphia, PA 19106, USA Tel +1 (800) 354 1420 Fax +1 (215) 625 2940

UK/Europe/Rest of World: The Advertising Manager, Routledge Journals, Taylor & Francis, 4 Park Square, Milton Park, Abingdon, Oxfordshire OX14 4RN, UK
Tel +44 (0) 207 017 6000 Fax +44 (0) 207 017 6336.

SUBSCRIPTIONS

Survival is published bi-monthly in February, April, June, August, October and December by Routledge Journals, an imprint of Taylor & Francis, an Informa Business.

Annual Subscription 2009

Institution	$520	£296	€435
Individual	$154	£90	€123
Online only	$494	£281	€413

A subscription to the institution print edition, ISSN 0039-6338, includes free access for any number of concurrent users across a local area network to the online edition, ISSN 1468-2699.

Dollar rates apply to subscribers in all countries except the UK and the Republic of Ireland where the pound sterling price applies. All subscriptions are payable in advance and all rates include postage. Journals are sent by air to the USA, Canada, Mexico, India, Japan and Australasia. Subscriptions are entered on an annual basis, i.e. January to December. Payment may be made by sterling cheque, dollar cheque, international money order, National Giro, or credit card (Amex, Visa, Mastercard).

Periodicals postage paid at Jamaica, NY by US Mailing Agent Air Business Ltd, c/o Priority Airfreight NY Ltd, 147–29 182nd Street, Jamaica, NY 11413.

US Postmaster: Please send address changes to Air Business Ltd, c/o Priority Airfreight NY Ltd, 147–29 182nd Street, Jamaica, NY 11413.

ORDERING INFORMATION

USA/Canada: Taylor & Francis Inc., Journals Department, 325 Chestnut Street, 8th Floor, Philadelphia, PA 19106, USA.

UK/Europe/Rest of World: Routledge Journals, T&F Customer Services, T&F Informa UK Ltd, Sheepen Place, Colchester, Essex CO3 3LP, UK.

Back issues: Taylor & Francis retains a three-year back issue stock of journals. Older volumes are held by our official stockists: Periodicals Service Company, 11 Main Street, Germantown, NY 12526, USA to whom all orders and enquiries should be addressed. *Tel* +1 518 537 4700 *Fax* +1 518 537 5899 *e-mail* psc@periodicals.com *web* http://www.periodicals.com/tandf.html

The issue date is February–March 2009.

The print edition of this journal is printed on ANSI conforming acid free paper by Bell & Bain, Glasgow, UK.

Closing Guantanamo: Is Europe Ready?

Sibylle Scheipers

President-elect Barack Obama's announcement that closing the US detention facility at Guantanamo Bay, Cuba, would be among his priorities has raised hopes among Europeans.[1] Reform of the detention system may be perceived as a first step towards the renewal of closer transatlantic ties, in line with the wider expectations of European governments and domestic publics towards the new US administration. Apart from the symbolic value, reform of US detention policy would also have the practical benefit of facilitating transatlantic cooperation in areas where it is most crucial: intelligence-sharing, transnational law enforcement and the conduct of multinational military operations.

European governments will also need to rethink their own approach to detention. Their reaction to US detention policies has encompassed both critical rhetoric and tacit acceptance (and sometimes assistance). These European governments have, however, not done enough to face up to the problem themselves.

Failure, blame and shame

The George W. Bush administration's spectacular mistakes regarding its detention policy after 11 September arose, according to critics, from a mixture of political and legal incompetence, strategic ignorance and moral laxity.[2] Although European governments disagreed with Washington on many aspects of the status and treatment of detainees in the so-called 'war

Sibylle Scheipers is Director of Studies, Oxford Leverhulme Programme on the Changing Character of War.

Survival | vol. 51 no 1 | February–March 2009 | pp. 5–12 DOI 10.1080/00396330902749608

on terror', that the topic developed into such a contentious and sensitive issue largely stifled attempts within and among European states to clarify their own approach to detention.

European engagement with the question of detention started rather late. Bush's announcement in early 2002 that terrorists and insurgents captured in Afghanistan would enjoy only the most minimal protections, if any, under the Geneva Conventions and customary international law led to widespread criticism in Europe. Yet European states contributing to the International Security Assistance Force (ISAF) in Afghanistan were slow to provide an alternative approach to detention. This changed only after allegations in early 2004 of ill-treatment and abuse of detainees in Guantanamo and in Abu Ghraib prison in Iraq. Europeans were faced with a dilemma: on the one hand, transferring detainees to US forces did not seem a viable option;[3] on the other, running detention facilities under European control did not appear to be feasible, not least due to limited troop numbers and resources. The eventual solution was to hand over responsibility for detention to the Afghan government. With ISAF's expansion (now under NATO command) to the whole of Afghanistan starting in 2005, NATO established an operational plan according to which detainees had to be transferred to the Afghan authorities within 96 hours of their arrest. Several European states subsequently concluded separate memoranda of understanding with the government of Afghanistan regulating the details of these transfers.[4]

The new policy was explicitly designed to avoid transfers of detainees from European forces to the United States – a stark illustration of the transatlantic gap in counter-terrorism policies that had developed since 11 September. It shifted the problem to the Afghan authorities, who at that time largely lacked the capabilities and infrastructure required to deal with this new task. Continuing reports about ill-treatment and disappearances of detainees in the Afghan prison system, release for bribes and frequent prison riots and prison breaks suggest that this situation has not changed much over the past few years. European NATO states (and Canada) could not agree on a single joint memorandum of understanding with the Afghan government. It appears that ideas on how to deal with the possibility that detainees could be transferred to third states via the Afghan authorities, and with the pitfalls related to the Afghan prison system, were too different.[5]

At the same time, allegations of the involvement of European states and their officials in US detention practices, first and foremost the so-called 'extraordinary renditions', became public. These ranged from allegations against Romania and Poland about harbouring secret detention facilities to other states' allowing stopovers of rendition planes on their territory.[6] While public scrutiny of this involvement is indispensable and should be regarded as proof of a healthy democratic debate, it had the side-effect of branding the issue of detention as predominantly associated with illegality and culpability. This, in turn, seems to have led many European policymakers and officials to shy away from tackling the topic up front. As a result, pragmatic opportunities to enhance the European approach to detention, such as capacity-building in Afghanistan, are often ignored.

From monitoring to capacity-building

It is a common and undisputed principle that states have a legal obligation to avoid transferring individuals in their custody to torture.[7] In addition, most European states regard their obligations under the European Convention for Human Rights as prohibiting transfers to situations where a detainee could face the death penalty. The memoranda of understanding concluded between European states and the government of Afghanistan include provisions to that effect. Most provide for monitoring mechanisms of transferred detainees in Afghan custody and for International Committee of the Red Cross access to these individuals. In addition, they require notification before a detainee can be transferred to a third state and stipulate that transferred individuals are not to be subjected to the death penalty.

These provisions have had mixed results. On the one hand, there have been allegations of torture of individuals transferred to Afghan custody. European officials became alert after the Canadian 'Afghan detainees scandal' broke in early 2007. The initial memorandum of understanding between Canada and the government of Afghanistan had not stipulated access to and monitoring of transferred detainees, and reports about abuse in Afghan custody emerged.[8] On the other hand, British forces suspended detainee transfers to the Afghan authorities after two previously transferred individuals seemed to have 'disappeared' in the Afghan prison system in 2007. Eventually, the disappearance turned out to be a bureaucratic mistake and the individuals were found.[9] That this incident resulted in a temporary suspension of detainee transfers shows that the monitoring system is working properly and that the British forces and officials are taking their legal obligations seriously.

Yet, in the wider context of prison reform in Afghanistan, the situation looks much bleaker. Apart from the risk of torture and disappearances, the Afghan prison system is beset by a plethora of problems stemming from insufficient capabilities and resources. For all the criticism of US detention policies, compared to European states Washington has proven more resourceful and determined in tackling the problem of Afghan prison reform. It has rebuilt large parts of Pul-e-Charki prison near Kabul, whereas European funding for such purposes has been desultory. The United Nations Office on Drugs and Crime (UNODC) has rehabilitated one block in Pul-e-Charki prison and a prison in Kabul with Italian funding. The UK has also provided funding for Pul-e-Charki. However, Afghan prison facilities are still notoriously overcrowded and far from meeting minimal standards. Frequent prison riots are the result.[10] The security of the prison facilities is poor as well: about half of all prison breaks worldwide in the past three years occurred in Afghanistan, with approximately 1,000 fugitives, half of them alleged to be involved in the insurgency.[11] A report about the largest prison break on 13 June 2008 at Sarposa prison near Kandahar concluded that the political sensitivities attached to the issue of detainees had inhibited efforts to improve the facility's security.[12]

Afghan prisons are notoriously overcrowded

Corruption and the release of prisoners for bribes is another widespread problem.[13] While corruption is endemic in many administrative branches in Afghanistan and will likely persist for many years, the large-scale failure to provide training for the Afghan National Police did nothing to decrease corruption in Afghan prisons. Initially, Germany was the lead nation for police reform, but handed this responsibility to the EU in 2007, having made no visible progress.[14] Police reform continues to be an area in which the multinational coalition in Afghanistan has not lived up to its promises, as has been reform of the judiciary, initially under Italian leadership.

Given that failures in police training and judiciary reform not only lead to poor detention conditions and corruption, but also increase the likelihood of ill-treatment in Afghan custody, it should be obvious that monitoring mechanisms are not sufficient to ensure that European states live up to their legal obligations regarding transfers to torture.[15] The more nervous governments have become, however, about allegations of torture in Afghan custody, the less emphasis they have placed on improvements in detention conditions. This means that Europeans are missing an important opportunity for long-term improvement of rule-of-law structures in Afghanistan. Due to its significant security presence, ISAF remains in a position to help the Afghan government, but once ISAF forces pull out, this window of opportunity will close.

Closing Guantanamo

European governments have not yet seized the opportunity afforded by the change of administration in Washington to clarify their expectations about detention reforms. Two requirements are particularly relevant: first, the Detainee Treatment Act, which in effect puts detainees held by the Department of Defence under Common Article 3 of the Geneva Conventions, should be extended to cover the CIA. Second, standards for security detention, both in theatres of military operations and on US territory, will need to be clarified. While security detention is legally permissible and has been practised by NATO forces in Kosovo, it should be conducted within a legal framework providing for adequate appeal and review procedures.[16]

The difficulties the Obama administration faces with respect to closing the detention facility at Guantanamo may provide an opportunity for Europeans to communicate these requirements while building a new foundation for transatlantic cooperation in the area of detention. Despite the fact that most European governments welcome plans to close Guantanamo, they have been remarkably silent with respect to the question of the fate of the individuals detained there. While some will face prosecution in the United States, most will have to be released or transferred, with the familiar problems of avoiding transfers to torture and, at least from the current US view, obtaining certain security assurances. So far only Albania and Sweden have accepted former Guantanamo detainees who are not their nationals.

Portugal's recent offer to resettle a number of Guantanamo detainees and its call on other European states to follow this example may be a model for others, but how other European governments will react remains to be seen.[17] Past experience indicates that their security concerns frequently outweigh their humanitarian motives. For instance, the German government refused to readmit former Guantanamo inmate Murat Kurnaz, a German resident with Turkish citizenship, onto German territory for several years before he was finally allowed to return in 2006. This reflects the stigma attached to Guantanamo inmates even if allegations against them fail to be substantiated after lengthy periods of detention.[18] By following the Portuguese example, European countries would increase the chances that Washington would at least listen to their requirements for future cooperation. However, if they insist Guantanamo is a problem that the United States should solve on its own, they miss the opportunity to explore the potential for future cooperation in an area that affects many aspects of counter-terrorism.

Acknowledgements

The author would like to thank Ashley Deeks for her helpful comments on earlier drafts of this article.

Notes

1 See, for instance, a letter from Portuguese Foreign Minister Luis Amado on the occasion of the 60th anniversary of the Universal Declaration of Human Rights, 10 December 2008, http://www.mne.gov.pt/mne/pt/noticias/200812101640.htm.

2 See, among others, Philippe Sands, *Torture Team: Deception, Cruelty and the Compromise of Law* (London: Allen Lane, 2008); Jane Mayer, *The Dark Side: The Inside Story of How the War on Terror Turned into a War on American Ideals* (New York: Doubleday, 2008).

3 It is unclear to what extent European forces in Afghanistan handed over detainees to the United States before 2004–05. Given that US forces were the only ones that had (and have) in-theatre detention facilities, at Bagram airbase and in Kandahar, it seems likely that transfers occurred. Such transfers were probably few, since the ISAF mandate was initially limited to Kabul and European states deployed relatively few troops under the Afghanistan-wide mandate of *Operation Enduring Freedom*.

4 Amnesty International, *Afghanistan – Detainees Transferred to Torture: ISAF Complicity?*, 13 November 2007, ASA 11/011/2007, available at http://www.unhcr.org/refworld/docid/473969ab2.html. The same applies to Canada. The US has reportedly also concluded a memorandum of understanding with the Afghan government.

5 Interviews with Legal Advisor, German Ministry of Defence, Bonn, March 2007, and with NATO Legal Advisor, Mons, January 2007.

6 Dick Marty, *Alleged Secret Detentions and Unlawful Inter-state Transfers involving Council of Europe Member States* (Strasbourg: Council of Europe, 2006). For the UK case in particular see the documents compiled by the All Party Parliamentary Group on Extraordinary Rendition, http://www.extraordinaryrendition.org/.

7 Relevant legal sources are the Convention Against Torture and, for European states, the European Convention on Human Rights.

8 Paul Koring, 'Amnesty Slams Canada over Afghan Detainees', *Globe and Mail*, 21 February 2007.

9 Phone interview with former Foreign and Commonwealth Office adviser in Helmand, December 2008.

10 See for example Kirk Semple, '8 Prisoners Die in Uprising at Afghan Prison', *International Herald Tribune*, 5 December 2008.

11 'Interpol pushes Afghanistan to record "terrorist" prisoners', Agence France-Press, 19 September 2008.

12 Nils N. French, 'The Sarposa Prison Break', *Canadian Army Journal*, vol. 11, no. 2, Summer 2008, pp. 8–15.

13 Sami Yousafzai and Ron Moreau, 'Afghan Prison Blues', *Newsweek*, 11 February 2008.

14 According to German General Hans-Christoph Ammon, head of the German Special Forces, this failure was due to insufficient funding and an inappropriate training approach. Judy Dempsey, 'German General Breaks Silence on Afghanistan', *International Herald Tribune*, 30 November 2008.

15 Poor police training is widely believed to increase the likelihood of torture and ill-treatment. The EU endorses the provision of effective police and judiciary training as one of the core instruments in its efforts to prevent torture in third countries. See *Guidelines to EU Policy towards Third Countries on Torture and Other Cruel, Inhuman or Degrading Treatment or Punishment* (Luxembourg: General Affairs Council, 9 April 2001). The Dutch forces seem to work on this assumption as well, since they are reported to transfer detainees only to those Afghan forces that they themselves have trained.

16 See Ashley Deeks, 'Administrative Detention in Armed Conflict', *Case Western Reserve Journal of International Law*, vol. 40, no. 3, 2008–2009 (forthcoming).

17 Letter from Portuguese Foreign Minister Luis Amado on the occasion of the 60th anniversary of the Universal Declaration of Human Rights.

18 See Laurel E. Fletcher and Eric Stover, *Guantanamo and Its Aftermath: U.S. Detention and Interrogation Practices and Their Impact on Former Detainees* (Berkeley, CA: Human Rights Center, 2008).

Muddling Through in Kosovo

Oisín Tansey and Dominik Zaum

Kosovo's unilateral declaration of independence of 17 February 2008 has made visible the deep divisions between the United States and its European allies on the one hand, and Russia on the other; divisions that shaped the political dynamics of the Kosovo crisis nine years ago as they do today. The failure to settle the status question through diplomacy has thrown the UN into crisis, leaving the Security Council deadlocked and the international community in Kosovo without direction and momentum. It has led to the de facto partition of Kosovo and control by Belgrade of the Serb-inhabited northern municipalities, and left the international community struggling to define the nature of its engagement. The political divisions that have heightened the problem in Kosovo over the last nine years are unlikely to be resolved soon and, if anything, recent developments have accentuated them. New and creative approaches to stabilising Kosovo and promoting its economic and institutional development are necessary. Current European Union projects in support of the peace process in Northern Ireland might offer a model for such engagement.

Oisín Tansey is Lecturer in International Relations at the University of Reading. He is the author of *Regime-Building: Democratization and International Administration* (Oxford University Press, forthcoming 2009). **Dominik Zaum** is Lecturer in International Relations at the University of Reading. His recent publications include *The Sovereignty Paradox: The Norms and Politics of International Statebuilding* (Oxford University Press, 2007); *The United Nations Security Council and War: The Evolution of Thought and Practice since 1945* (Oxford University Press, 2008), co-edited with Vaughan Lowe, Jennifer Welsh and Adam Roberts; and (with Adam Roberts) the Adelphi Paper *Selective Security: War and the United Nations Security Council since 1945* (Routledge, 2008).

Survival | vol. 51 no. 1 | February–March 2009 | pp 13–20 DOI 10.1080/00396330902749632

International context

The divisions between Kosovo Serbs and Albanians in Kosovo are increasingly mirrored by a divide within the international community, reflected especially in the diplomatic response to the unilateral declaration of independence. Immediately after the announcement Kosovo was recognised by the United States and most members of the EU, and by December 2008 53 countries, including four of the six former republics of Yugoslavia, had extended recognition. However, a majority of states have yet to respond positively and many are actively opposed to the move. Moscow has lobbied states extensively to refuse recognition, arguing that the move to independence is a breach of international law that endangers international stability. Pristina has been particularly disappointed that only a small number of states from the Organisation of the Islamic Conference have recognised Muslim-majority Kosovo, a pattern than can in part be explained by Moscow's lobbying efforts and political pressure. Several EU countries have also objected to Kosovo's independence and refused to recognise the new state. Spain has led the opposition within the EU, not least due to concerns about the precedent it might set for its own territorial integrity, and has lobbied Latin American states in particular to withhold recognition.

Russia's expressed concerns that Kosovo's independence could lead to further changes in the international order became something of a self-fulfilling prophecy after its conflict with Georgia in August 2008. Citing the recognition of Kosovo as a precedent when extending its own recognition to Georgia's two pro-Russian breakaway entities, Russia argued that the circumstances in South Ossetia and Abkhazia represented a 'special situation' similar to the one Western countries had used to justify the legality of Kosovo's independence.[1] This move, however, undermined Russia's argument in relation to Kosovo that territorial integrity was sacrosanct, and led to concerns in Belgrade that it might weaken the case against Kosovo's independence.

A settlement of the Kosovo question has been further complicated by Serbia's diplomatic efforts to limit international support for independence. Since Serbian elections in May 2008 the new government in Belgrade has taken a more moderate stance on Kosovo and has been more open to cooperation with the EU presence. Belgrade also prevailed on the UN General Assembly to request an advisory opinion from the International Court of Justice on the legality of Kosovo's declaration.[2] Should the court consider the declaration of independence a breach of international law, it seems unlikely that states that have already recognised Kosovo will withdraw recognition, but it might further slow the trickle of recognitions. It might lead Serbia to accept that, though it can now claim the moral high ground, it needs to move on and engage pragmatically with a Kosovo whose independence is irreversible. At worst, such a ruling would deepen political uncertainty in Kosovo and the region, and hamper its political and economic development.

The plan put forward by UN Special Representative Maarti Ahtisaari in 2007 envisaged that independence for Kosovo would be supervised by two new EU-led missions, an International Civilian Office (ICO) and a European Union rule-of-law mission (EULEX).[3] The Security Council failed to endorse the Ahtisaari Plan, and its implementation has been deeply problematic. While EULEX and the ICO were established and deployed to Kosovo following the declaration of independence, Russia has blocked in the Security Council the termination of the UN Mission in Kosovo (UNMIK), as this would imply the recognition of Kosovo's independence.[4] Moscow has challenged the authority of the new EU missions, and without the backing of a UN Security Council mandate both the ICO and EULEX have struggled to carve out clear roles or establish effective authority in Kosovo. An agreement in November 2008 to reconfigure the international presence to include EULEX under a status-neutral 'UN umbrella' has been opposed by Kosovo Albanians and led to protests against EULEX in Pristina, and merely papers over the continued deep divisions in the Security Council.[5]

Local developments

With the exception of the destruction of two border posts in the north by Kosovo Serb protesters in the days after the declaration of independence, and clashes between UNMIK police, NATO's Kosovo Force (KFOR) and Serbs around the north Mitrovica courthouse on 17 March (which left an UNMIK policeman dead), Kosovo has largely been peaceful since declaring independence. However, the international divisions over its recognition and the future of UNMIK have contributed to a leadership vacuum in the international presence and have further entrenched the division between Kosovo's institutions and the Serb community.

The size and diversity of the international presence, and the scale of the challenge of integrating and strengthening Kosovo's political institutions, underline the importance of clear leadership to ensure coherence of international policy in Kosovo and to maintain the momentum of peacebuilding. Currently none of the missions provide such leadership. UNMIK has no legitimacy with the Albanian majority; as the mission's head, Lamberto Zannier, noted, 'attempts to impose my legal authority are simply not heeded by the Kosovo Albanian majority, which now sees the Constitution of Kosovo as the fundamental document from which legal authority derives'.[6]

The new international presence envisaged in the Ahtisaari Plan has not been able to fill this gap. With two new and separate offices, which have overlapping responsibilities but separate chains of command, its structure is inherently complex, and neither the International Civilian Office nor EULEX have undisputed primacy. The problem is made worse by the legal uncertainty about their authority arising from the inability to close down the UN mission and fully transfer power. The secretary-general's six-point plan to facilitate the reconfiguration of UNMIK and the transition to the EU in Kosovo remains creatively ambiguous about the authority of EULEX, and even more so about the role of the International Civilian Office, which is not mentioned at all. EULEX emphasises that it is a technical rule-of-law mission, which inherently limits the wider political leadership it could provide. It is also constrained by the continued divisions within Europe over Kosovo's independence. While the International Civilian Office has worked closely with the Kosovo government to implement the Ahtisaari Plan's legislative agenda, it has been unable to establish a strong working relationship with the other main international actors, such as UNMIK or the Organisation for Security and Cooperation in Europe, which would allow it to coordinate their efforts and provide political leadership.[7]

One of the most important consequences of this vacuum has been an inability to reverse the deepening de facto partition between the territories inhabited by Kosovo Serbs and the rest of Kosovo. The authority of the

Kosovo government is contested in all Serb municipalities, where Serbs have set up new municipal administrations, financially supported by Belgrade, that have not been recognised by either the government or the international presence in Kosovo. In northern Mitrovica and the three municipalities north of the river Ibar, Albanian members of the police have been withdrawn, and the remaining Kosovo Serb police are largely controlled by the local presence of the Serbian Ministry of the Interior. EULEX was not present in the north until mid December, and the International Civilian Office still has no offices there. Smuggling and organised crime are thriving in the border areas.

> *Kosovo government authority is contested in all Serb municipalities*

The political economy of northern Kosovo may pose the central challenge to greater stability in Kosovo, and addressing it should be at the core of the international strategies to move beyond the current impasse. Kosovo Serbs in the north are highly dependent on financial subsidies from Belgrade: research by the European Stability Initiative suggests that in 2003, more than 60% of all cash income in northern Mitrovica came from salaries and social transfers paid by the Serb government – approximately €1.6 million per month.[8] The Kosovo government does not have the funds to provide a strong economic incentive for Kosovo Serbs to participate in post-independence political and social institutions. Members of the international community, in particular European states and institutions, have the necessary resources but have not used them, hampered by an unwillingness to engage with the unrecognised Serb municipal administrations.

Restoring momentum

To regain momentum in Kosovo, the EU should focus on its financial muscle and its ability to encourage cooperation on apparently technical issues to elicit greater political integration. Opinion polls show that poverty and unemployment are the two greatest concerns of Kosovo's population, independent of ethnic group.[9] This creates an opportunity for the EU to use well-targeted financial aid to loosen the grip that Belgrade has in northern

Kosovo and the enclaves, and to facilitate greater engagement of Kosovo Serbs with Kosovo's political institutions. The EU could explore a strategy it has already tried with success in Northern Ireland, where 'peace money' has played a role in encouraging and facilitating inter-group cooperation.

The EU's Special Support Programme for Peace and Reconciliation has directed over €1 billion into Northern Ireland in targeted grants, often requiring joint inter-community applications. Evidence suggests that it has had significant impact on inter-community relations.[10] The principal current EU mechanism for directing money into Kosovo is the Instrument for Pre-Accession Assistance, a programme that governs EU aid to all candidate and potential candidate members in the region. The programme for Kosovo includes projects for capacity building and the rule of law, as well as post-conflict issues such as refugee returns and reintegration. Yet Kosovo would benefit further from a dedicated peace and reconciliation programme designed to overcome the intense social and economic divisions within the territory.[11] Such a programme could involve allocation of funds at the local level through inclusive cross-community funding bodies, and could make economic assistance for joint projects that include members from across the community divide a priority – the core strategies of the Northern Ireland Programme. Not only would local funding boards help promote greater interaction, they would help channel funds in ways that avoid Pristina-based corruption. This would provide economic incentives for cooperation in areas where communities are currently highly segregated, and could lessen the economic and political leverage that Belgrade currently enjoys in the north. With some creative strategic thinking, the EU has the potential to fill the current leadership vacuum and significantly improve conditions on the ground. Without it, the sense of political and economic stagnation will create frustration and further instability.

Notes

1 'Russia Backs the Independence of Georgian Enclaves', *New York Times*, 26 August 2008. See also Oksana Antonenko, 'A War with no Winners', *Survival*, vol. 50, no. 5, October–November 2008, pp. 23–36.

2 United Nations, 'Request for an Advisory Opinion of the

International Court of Justice on whether the Unilateral Declaration of Independence of Kosovo is in Accordance with International Law', resolution adopted by the General Assembly, A/RES/63/3, 8 October 2008, available at http://daccessdds.un.org/doc/UNDOC/GEN/N08/470/97/PDF/N0847097.pdf.

3 UN Security Council, *Report of the Special Envoy of the Secretary-General on Kosovo's Future Status*, addendum to letter dated 26 March 2007 from the secretary-general addressed to the president of the Security Council, S/2007/168/Add.1, 26 March 2007, available at http://daccessdds.un.org/doc/UNDOC/GEN/N07/279/01/IMG/N0727901.pdf.

4 UN Security Council Resolution 1244, S/RES/1244, 10 June 1999, available at http://www.unmikonline.org/press/reports/N9917289.pdf.

5 UN Security Council, *Report of the Secretary-General on the United Nations Interim Administration Mission in Kosovo*, S/2008/692, 24 November 2008, available at http://www.unhcr.org/refworld/docid/492e79152.

6 Minutes of the 5944th meeting of the UN Security Council (provisional), 25 July 2008, S/PV.5944, p. 2, available at http://www.securitycouncilreport.org/atf/cf/{65BFCF9B-6D27-4E9C-8CD3-CF6E4FF96FF9}/Kos SPV 5944.pdf.

7 International Crisis Group, *Kosovo's Fragile Transition*, Europe Report no. 196 (Brussels and Pristina: ICG, 2008), p. 9, available at http://www.crisisgroup.org/home/index.cfm?id=5695.

8 European Stability Initiative, *A Post-Industrial Future? Economy and Society in Mitrovica and Zvecan* (Berlin and Pristina: ESI, 2004), p. 6. See also Kosovo Institute for Policy Research and Development, 'Looking Beyond Mitrovica Bridge: An "Ahtisaari Plus" Package Proposal', Policy Brief Series no. 9, October 2008, available at http://www.kipred.net/site/documents/PB_09_ENG.pdf.

9 UN Development Programme, *Early Warning Report Kosovo 20/21: January–June 2008* (Pristina: UNDP, 2008), p. 27.

10 For details of the EU programme, see European Commission, *EU Programme For Peace and Reconciliation In Northern Ireland and The Border Region of Ireland, 2000–2004, Operational Programme*, CCI No. 2000.RG.16.1.PO.001, available at http://ec.europa.eu/regional_policy/country/overmap/pdf_region/fp2mc_en.pdf. For an assessment, see Cathal McCall and Arthur Williamson, 'Fledgling Social Partnership in the Irish Border Region: European Union "Community Initiatives" and the Voluntary Sector', *Policy & Politics*, vol. 28, no. 3, July 2000, pp. 397–410.

11 For the importance of economic policy supporting political integration in post-conflict situations, see Graciana del Castillo, *Rebuilding War-Torn States: The Challenge of Post-Conflict Economic Reconstruction* (Oxford: Oxford University Press, 2008).

Dangerous Waters

Ken Menkhaus

The epidemic of piracy off the coast of Somalia since 2007 included the spectacular pirating in November 2008 of the Saudi supertanker *Sirius Star* containing a quarter of Saudi Arabia's daily oil production; the ship was released on 10 January 2009 in exchange for a $3 million ransom. The epidemic has led to extensive media commentary, a flurry of UN Security Council resolutions, and deployment of an unusual combination of some of the world's most powerful navies to escort commercial ships through one of the world's busiest, and now most dangerous, shipping lanes.

Predictably, policymakers, pundits and politicians have sought to harness the piracy story to advance their own agendas. Many have seized on it to plead for a durable political solution to the 19-year crisis of state collapse in Somalia. This is a waterborne variation on the 'securitisation' of state-building, the argument that failed states pose a host of spillover dangers, including piracy, if left unresolved. Humanitarians have contrasted the robust international anti-piracy response with the tepid global mobilisation to address Somalia's horrific humanitarian crisis, in which three million Somalis are in need of emergency aid. Some apologists have cast Somali piracy in 'Robin Hood' terms, as a legitimate local response by poor coastal fishing communities against external predators engaged in illegal fishing and toxic-waste dumping in their waters. More than a few journalists have framed the Somali piracy story as another example of 'Mad Max anarchy'

Ken Menkhaus is Professor of Political Science at Davidson College, and a specialist on the Horn of Africa. His recent publications include *Somalia: State Collapse and the Threat of Terrorism* (2004) and 'Governance without Government in Somalia' in *International Security* (2007).

Survival | vol. 51 no. 1 | February–March 2009 | pp. 21–25 DOI 10.1080/00396330902749640

prevailing in the hopelessly ungovernable place called Somalia. Still others warn that it is only a matter of time before the pirates fire on a ship laden with chemicals or fuel, producing an environmental disaster on Somalia's shoreline. And counter-terrorism experts have expressed concern over a 'terrorism–piracy nexus' in which some of the tens of millions of dollars of ransom money flowing into Somalia are being garnered by the al-Qaeda affiliate in Somalia known as al-Shabaab. What all of these perspectives share is the claim that Somali piracy is only a symptom of a much bigger political problem on land.

This is emphatically true – the Somali piracy epidemic is unquestionably an on-shore crisis demanding an on-shore solution. Naval operations to interdict and apprehend pirates will help, but cannot possibly halt the daily quest of over a thousand gunmen in such vast waters when the risks are so low, rewards so high and alternatives so bleak in desolate Somalia. As for the other narratives, there is at least some truth to all them, but also much that is potentially misleading.

The coast of Somalia has been the world's worst piracy area only since 2007. But piracy and other variations on extortion and kidnapping at sea have been at play in Somalia since the early 1990s, when the state first collapsed and warlords sought new ways to parley their firepower into profit. In 1991, foreign fishing trawlers aggressively moved into Somalia's rich and unpatrolled waters, at the expense of coastal fishing villages. Angry Somali fishermen secured weapons and began firing on foreign trawlers. But what began as a genuine grievance on the part of local fishermen quickly morphed into an enterprise by militia leaders, who saw an opportunity to earn easy cash by demanding fees from foreign fishing ships for 'licences' to fish in waters near territory they controlled. Those who failed to pay the fees ran the risk of capture and kidnapping of the crew. The militia leaders overseeing these 'patrols' were utterly unconcerned about the inhabitants of beleaguered fishing communities, who are generally treated as second-class citizens in Somalia. Nor were militia leaders especially aggrieved at the illegal foreign harvesting of Somalia's fisheries. They were simply seizing upon an easy form of extra income, and were in some ways complicit in the foreign ransacking of Somali fisheries. The Robin Hood narrative of Somali piracy as a grassroots form of coastal patrol against rapacious foreign fishing vessels is thus only partly true, and at any rate has long since been overtaken by less noble motives. For scholars exploring war economies, Somali piracy is a textbook case of a shift in the motives of an armed group from grievance to greed.

What is puzzling is why it took so long for land-based criminal gangs and militias to realise the enormous potential of piracy just off their shores. The act of piracy is little more than an extension of activities that armed groups have engaged in for years: militia roadblocks, extortion and kidnapping for ransom are a staple source of income for gangs and militias in Somalia. The pirates do employ a few new technologies, such as GPS devices, cell phones, Thuraya (satellite) phones, and advanced money-counting machines to verify that payments are in full and not counterfeit. But the basic act of piracy is surprisingly low-tech, and could have been achieved at any point since 1991.

Small groups, including some former fishermen, some militiamen armed with AK-47s and rocket-propelled grenades, and at least one technically savvy operative, use small speedboats to capture a dhow or fishing vessel, then use that as a mother ship to ply waters far from shore in search of unguarded, lightly manned, slow-moving cargo ships. Boarding the ships is often accomplished within 15 minutes and without a shot, at which point the ship is ordered to sail for one of several pirate villages along Somalia's northeastern coast. From there, the negotiators take over, bargaining via cell phone over the course of weeks for ransoms from $500,000 to over one million dollars. Ransom for the ship and crew is the only objective of the pirates. To date, they have shown no inclination to offload contents of the ships, steal the vessels or use the captured ships for acts of terrorism. Ship owners have been willing to pay the ransoms because the value of the ship and crew are so much greater.

Ransom for the ship and crew is the only objective

It is estimated that $20–$40m has been paid in ransom for the 60 or so ships pirated in 2008 alone. Put into perspective, this is a telling figure. On the one hand, it is a paltry sum, a 'nuisance tax' for the insurance companies of the more than 16,000 ships that pass through the Gulf of Aden each year. The ransom money adds up to only 3% of the $900m sought by the UN for emergency humanitarian aid for Somalia in 2009. The total costs of operating over two dozen warships off the Somali coast to protect shipping from piracy is unknown but certainly exceeds the total ransom paid many times over.

On the other hand, $40m is a transformative amount of money inside impoverished Somalia, a country which typically earns no more than $100m annually from its main export, livestock. Indeed, the funds accrued by pirates equal the annual budget of the unrecognised secessionist state of Somaliland. Not surprisingly, this level of easy money has quickly attracted attention in Somalia, leading to the rise of a mafioso racket in which top political and business figures are earning sizable cuts of the ransoms. It is likely that at least some of the funds find their way into the hands of the jihadist al-Shabaab group as well. The fact that the ransom money touches so many hands in northeastern and central Somalia creates a serious disincentive on the part of the leaders of those regions to address the problem. Indeed, the worry now is that the autonomous state of Puntland in northeast Somalia is becoming a pirate version of a narco-state.

From a strictly economic perspective, a case could be made that the international community is better off doing nothing, and treating the ransoms as the price of doing business through the Gulf of Aden. But Somalia's piracy epidemic is not just about the cost of ransoms. It is a serious threat to Somali commerce, especially the flow of humanitarian food shipments, and is harming regional commerce from Kenya to Djibouti. The biggest worry for the international community is the prospect that Somalia's increasingly powerful, radical al-Shabaab movement will take control of the piracy and harness it for terrorist purposes, either to raise money for their cause or to use captured ships in a terrorist attack. The nightmare scenario is that of a 'rogue supertanker' driven into a regional seaport. To be sure, it is not at all clear that, if given the chance, al-Shabaab will try to use piracy for political purposes. On the contrary, in 2006 Somali Islamists who briefly gained control of south-central Somalia succeeded in shutting the pirates down, and al-Shabaab is likely to calculate that putting an end to piracy again will enhance their credibility in the eyes of Somalis and the world. But with much greater flows of money coming in by ransoms, the temptation to exploit rather than stop piracy will be considerable. As al-Shabaab advances on some of the pirate villages, we will soon learn what its policies on piracy will be.

Meanwhile, recent UN Security Council resolutions have helped to resolve a number of legal impediments to interdiction of piracy. The Security Council authorises foreign navies to enter Somali territorial waters in pursuit of pirates and authorises them to attack on shore as well. The UN also encourages bilateral arrangements with regional states to accept captured pirates for trial before a court of law. Prior to this, naval patrols which captured Somali pirates often released them on shore because they had no legal alternatives. What the UN Security Council has not done, however, is to create a coordination mechanism for the dozen or more navies agreeing to patrol the Gulf of Aden and provide escorts for commercial ships. This will be a fascinating issue to observe, as the navies involved include those from the United States, European Union, China, Iran, Malaysia and India.

As the international community contemplates what to do beyond the stopgap naval patrols, one cautionary note must be kept in mind. The value of helping train and supply a Somali coast guard under the control of regional authorities seems self-evident, and under the right conditions is advisable. Yet some of the pirates are themselves ex-coast guard. The knowledge imparted in training up a coast guard can easily become a valuable skill-set for pirates. As long as criminality is much more lucrative and lower risk in Somalia than law enforcement, efforts to strengthen local police and security-sector capacity run the risk of strengthening the criminals instead. Ethiopia, which is now withdrawing its military forces from southern Somalia after a failed two-year occupation, learned this the hard way when it watched in frustration as most of the Somali security forces it trained immediately defected to the Islamist insurgency. Whatever the international community opts to do in response to Somali piracy, it must bear in mind that in Somalia the law of unintended consequences is not just a theory. It's the law.

Noteworthy

Mumbai attacks

'We have to galvanise the international community into dealing sternly and effectively with the epicentre of terrorism, which is located in Pakistan.'

Dr Sanjaya Baru, former official spokesman and media adviser to the prime minister of India, outlines the steps he believes must be taken to end terror attacks like those carried out in Mumbai[1]

'To say that Pakistan is the epicentre of terrorism is highly unfair. The internal dynamics of India encourage such kind of terrorism.'

Dr Jafar Iqbal Cheema, senior adviser on arms control and nuclear non-proliferation at the South Asian Strategic Stability Institute, insists that Pakistan is not the only state struggling to control the activities of terrorist groups[2]

'There does need to be action. There needs to be action urgently and transparently. And it's a message that has been well received in Pakistan, because it's Pakistan's fight as well.'

US Secretary of State Condoleezza Rice comments on the need to control terrorist organisations operating from Pakistan, following a meeting with Pakistani President Asif Zardari[3]

Half full

'I think, given the circumstances we've had to deal with, we've done pretty well.'

Outgoing US Vice President Dick Cheney assesses the accomplishments of the Bush administration[6]

The elect

'If a foreign enemy attacks the United States during the Harvard–Yale game any time over the next four years, we're screwed.'

David Brooks comments on the super-abundance of the best and the brightest in Barack Obama's administration[7]

1,800,000

Number of pints of beer shipped to German troops in Afghanistan in 2007[4]

Foreclosure

'We could claim Alaska – it was only granted on lease, after all.'

Professor Igor Panarin, a Russian political analyst, predicts that a declining American economy will cause the United States to break up[8]

Interesting times

'I wouldn't say that it's a good thing … for a small country – we'd rather prefer to have an uninteresting, boring life.'

Eka Tkeshelashvili, Georgian minister for foreign affairs, comments on Georgia's higher international profile following its August war with Russia during a speech at the IISS in London, 26 November 2008[5]

A woman's place

'I think that there probably, someday again will be a male secretary of state.'

Former US Secretary of State Madeleine Albright comments on the diplomatic landscape at the FORTUNE 500 Forum, 1 December 2008[9]

Survival | vol. 51 no. 1 | February–March 2009 | pp. 26–28

DOI 10.1080/00396330902749665

Manama Dialogue 2008

'We cannot make every country in the world look like Vermont.'

US Secretary of Defense Robert Gates[10]

'In case anyone else neglects to say it, thank you, the United States.'

Charles Powell, former foreign-policy adviser to UK Prime Minister Margaret Thatcher and member of the IISS Council, thanks the United States for its efforts to return stability to Iraq[11]

'The new Iraq is not perfect ... If there is another country in this region that is perfect, please let us know so that we can emulate it.'

Dr Barham Saleh, deputy prime minister of Iraq[12]

'We do not want to be an empire, and we are not a "second edition" of the Soviet Union. We do not believe in revolutions anymore. We have had enough of those. We believe in evolutions.'

Mikhail Margelov, chairman of Russia's Committee for Foreign Affairs[13]

'Some journalists report a nonsense that 72% of Afghans are under Taliban control; if that were the case, I would not be able to travel and talk to you today.'

Dr Zalmai Rassoul, Afghanistan's national security adviser[14]

Cold call

'They've gotten Hugo Chávez and others to fall for their tricks. I said, "Oh, no, I won't be punked".'

US Congresswoman Ileana Ros-Lehtinen explains why she hung up on President-elect Barack Obama, thinking the call was a radio hoax.[15]

$100m

Value of the oil on board the tanker *Sirius Star* hijacked by Somali pirates in November 2008[16]

$3m

Ransom paid for the release of the ship, cargo and crew in January[17]

Sources: 1–2 The 5th IISS Regional Security Dialogue, The Manama Dialogue, 12–14 December 2008, http://www iiss org/conferences/the-iiss-regional-security-summit/, **3** 'Rice Pakistan Knows it Has to Act Urgently', CNN, 4 December 2008, http://edition cnn com/2008/WORLD/asiapcf/12/04/india pakistan rice/index html, **4** Kate Connolly, 'Troops' Beer Allowance a Headache for Germans', *Guardian*, 15 November 2008, http://www guardian co uk/world/2008/nov/15/germany-afghanistan-beer, **5** Eka Tkeshelashvili, address to the IISS, London, 26 November 2008, http://www iiss org/recent-key-addresses/eka-tkeshelashvili-address/, **6** Ed Pilkington, 'Dick Cheney has No Regrets About Eight Years in Office', *Guardian*, 17 December 2008, http://www guardian co uk/world/2008/dec/17/dick-cheney-interview-guantanamo-waterboarding, **7** David Brooks, 'The Insider's Crusade', *New York Times*, 21 November 2008, http://www nytimes com/2008/11/21/opinion/21brooks html?_r=2&em, **8** 'Russian Analyst Predicts Decline and Breakup of U S', Russian News and Information Agency, 24 November 2008, http://en rian ru/world/20081124/118512713 html, **9** Quoted in Mike Allen, '49 Days to Inauguration', Politico, 2 December 2008, 'http://www politico com/playbook/1208/playbook511 html, **10–14** Manama Dialogue, **15** 'Congresswoman Hangs Up on Obama', BBC, 4 December 2008, http://news bbc co uk/2/hi/also_in_the_news/7765574 st, **16** Xan Rice and Matthew Weaver, 'Sirius Star Pirates Demand $25m Ransom', *Guardian*, 20 November 2008, http://www guardian co uk/world/2008/nov/20/piracy-somalia, **17** Mohammed Ibrahim, 'Pirates Say They Freed Saudi Tanker for $3 Million', *New York Times*, 9 January 2009, http://www nytimes com/2009/01/10/world/africa/10somalia html?_r=1&ref=world

KYRGYZSTAN
CHINA
UZBEKISTAN
TAJIKISTAN
TURKMENISTAN
JAWZJAN
BALKH
KUNDUZ
BADAKSHAN
TAKHAR
SAMANGAN
BAGHLAN
NORTH WEST FRONTIER
FARYAB
SARI PUL
PANJSHER
NURISTAN
KUNAR
KAPISA
PARWAN
LAGHMAN
BADGHIS
BAMIYAN
Kabul
Jalalabad
NANGARHAR
AFGHANISTAN
WARDAK
Peshawar
KABUL
LOGAR
HERAT
GHOR
HAZARAJAT
PAKTIA
TRIBAL AREA
Islamabad
KHOWST
Khowst
ORUZGAN
GHAZNI
FARAH
PAKTIKA
WAZIRISTAN
ZABOL
Farah
Kandahar
PUNJAB
NIMROZ
HELMAND
KANDAHAR
PAKISTAN
IRAN
BALUCHISTAN
INDIA
Miles 120
Km 200
© IISS

Doctrine and Reality in Afghanistan

Adam Roberts

The limitations of military doctrines and practice are often exposed, not by arguments, but by events. Thus it was mainly events in Iraq and Afghanistan that exposed the inadequacies of the so-called 'revolution in military affairs' – an idea that was popular in the United States from the mid 1990s until at least 2003.[1] Now, Afghanistan – and the situation in Pakistan with which it is inextricably linked – is proving to be a harsh test of the revived ideas of counter-insurgency.

Afghanistan was always likely to be a difficult theatre of operations for outside military forces. Seeing this (and perhaps also because he did not want an ongoing distraction from the future invasion of Iraq, for which he was already lobbying), then-US Deputy Secretary of Defense Paul Wolfowitz said in November 2001:

> In fact, one of the lessons of Afghanistan's history, which we've tried to apply in this campaign, is if you're a foreigner, try not to go in. If you go in, don't stay too long, because they don't tend to like any foreigners who stay too long.[2]

The wars in Afghanistan in the nineteenth century have been the foundation for a view of the country and its peoples – especially the latter – as

Adam Roberts is President-elect of the British Academy and Senior Research Fellow at the Centre for International Studies in Oxford University's Department of Politics and International Relations. He is also an Emeritus Fellow of Balliol College, Oxford. He was the Montague Burton Professor of International Relations at Oxford University from 1986 to the end of 2007.

Survival | vol. 51 no. 1 | February–March 2009 | pp. 29–60 DOI 10.1080/00396330902749673

unusually resistant to any kind of foreign influence or control, actual or perceived. David Loyn, the veteran BBC reporter on Afghanistan who has charted these previous conflicts, argues that mistakes are being repeated today because of a neglect of the study of history. He charges that the United States and Britain have failed to understand the extent of resistance in Afghanistan to anything that looks like foreign control.[3]

The war in 1979–89 between the Soviet-backed government of Afghanistan and its mujahadeen adversaries contributed to the collapse of the Soviet empire – not only proof of the fateful consequences that may flow from war in Afghanistan, but also one driver of the present war. The channelling of much international aid to mujahadeen groups through Pakistan reinforced the fateful link between events in Pakistan and those in Afghanistan. The power of non-state groups and regional military chiefs, and their tendency to rely on threats and uses of force not controlled by any state, became more deeply ingrained than before in both Afghanistan and the frontier areas of Pakistan. The religious element in Afghan politics did not disappear with the departure of Soviet forces in 1989. Indeed, within a few years religious warriors trained in the hard school of combat against Soviet forces in Afghanistan were to turn up in a wide range of other locations, including in the former Yugoslavia. These legacies of the war against Soviet control remain most important in Afghanistan itself. The problems of non-state violence, regional rivalries and the religious element in politics are not new to Afghanistan, but they were reinforced. Above all, the old instinctive suspicion of foreign projects is still there.

Following the withdrawal of the last Soviet forces from Afghanistan in January 1989, there was an internal crisis and civil war that has never really ended. The first phase was only partially concluded on 26 September 1996 when Kabul fell to the Taliban, which established a theocratic style of government throughout the areas under their control. The United Islamic Front for the Salvation of Afghanistan, or Northern Alliance, continued to control an area of northern Afghanistan and to challenge Taliban rule. Following the 11 September 2001 al-Qaeda attacks on the United States, direct US and coalition military intervention in Afghanistan changed the character of this continuing war. It became, for a few months, a war between sovereign states

– the US-led coalition versus the Taliban government of Afghanistan. In November–December 2001 the US-led intervention, and the military campaign of the Northern Alliance, toppled the Taliban regime, which had been supported by al-Qaeda. Initially there was much enthusiasm in Kabul and elsewhere for the incoming International Security Assistance Force (ISAF), but this situation was to change.

The international war in 2001 had been superimposed on two more enduring conflicts: between the Taliban and the Northern Alliance, and between the United States and its allies against al-Qaeda terrorists. Both continued. The war against al-Qaeda and related terrorists, now based in Pakistan as well as Afghanistan, carried on without interruption. In addition, there was growing resistance in southern Afghanistan to the new regime. This insurgency began relatively slowly, so that its seriousness was not recognised for some time. It is commonly labelled the 'Taliban insurgency' – a description that may conceal the possibility that the sources of support for the insurgency have been more numerous than the label 'Taliban' suggests, or that the ideology of the Taliban may have evolved. The insurgent movement has drawn on elements of both Afghan and Pashtun nationalism; it has operated alongside traditional forms of social organisation and systems of justice; its recruiting has been facilitated by Afghanistan's high levels of unemployment and by the fact that it is able to pay its soldiers good money; and its willingness to support poppy cultivation adds to its support in certain provinces while exposing the incoherence of the various NATO countries on this issue.[4] None of this is to suggest that all those forces labelled 'Taliban' should be seen simply as heroic patriots or as Pashtun traditionalists. Ahmed Rashid has written:

> The United States and NATO have failed to understand that the Taliban belong to neither Afghanistan nor Pakistan, but are a lumpen population, the product of refugee camps, militarised madrassas, and the lack of opportunities in the borderland of Pakistan and Afghanistan. They have neither been true citizens of either country nor experienced traditional Pashtun tribal society. The longer the war goes on, the more deeply rooted and widespread the Taliban and their transnational milieu will become.[5]

Into this ongoing conflict a new element was added from 2005: the involvement in combat activities of contingents of the NATO-led ISAF. Initially the UK had taken the lead in organising ISAF in January 2002, followed at six-monthly intervals by other 'lead states' until NATO as such took over in August 2003. ISAF's remit gradually extended across Afghanistan, and in some provinces came to involve direct combat.[6] By 2006 ISAF comprised troops from 32 countries. Those deployed in the southern provinces of Afghanistan became increasingly geared to a counter-insurgency campaign, the transition to which resulted in uneven burden-sharing between NATO member states. NATO had put itself in the unenviable position of staking its impressive reputation on the outcome of a distant and little-understood war in a country well known to be a graveyard for foreign military adventures.

The involvement of outsiders has one further special characteristic: it is based on short-term tours of duty, so there is remarkably limited institutional memory, especially as regards knowledge of local communities and political traditions. Few outsiders involved in international civil and military work in Afghanistan have learned the relevant languages.

There is remarkably limited institutional memory

One feature of the ongoing war in Afghanistan that distinguishes it from some other post-Cold War US involvements has been that the US-led forces had, at the start, significant allies within the country: originally the Northern Alliance, then the government of Afghanistan. This made the Afghan involvement different from some of the other conflicts in which the United States has been involved, including Iraq in the first years of the US-led presence and Somalia over a much longer period. But this apparently favourable situation had inherent limitations and was vulnerable to change. The Northern Alliance, even at the best of times an unstable coalition, never controlled all of Afghanistan. The Afghan authorities conspicuously lacked the bureaucratic back-up that provides the essential underpinning of most governments around the world. The Pashtuns generally resented the Northern Alliance's US-assisted victory in December 2001; and when, around 2003–04, the Pashtuns came back strongly in the government (thanks to the new constitution and law on political parties), Afghan opinion critical of the United States found a voice. Indeed, the boot was now on the other foot, with minorities complaining of Pashtun nationalism and structural exclusion.

Afghanistan's neighbours – including China, Iran, Tajikistan, Turkmenistan and Uzbekistan – all have legitimate interests in the country and its long-running conflicts. Many other states, including India and Russia, also have legitimate interests in whether Afghanistan can manage to stay together, make progress in development, and attract refugees back. The relationship with Pakistan is the most complex, and has contributed most to Afghanistan's ongoing divisions. The Pakistan connection has deeply affected events in Afghanistan in all the wars there since the Soviet intervention in 1979. Throughout, Pakistan's Inter-Services Intelligence (ISI) has had a major, and not always controlled, role. In the 1980s Pakistan, with massive Western support, provided crucial assistance for the anti-Soviet rebels in Afghanistan. Then from 1994 onwards there was extensive Pakistani official support for the Taliban movement in Afghanistan.[7]

In the ongoing war in Afghanistan a number of consequences have flowed from the Pakistan connection. The first is that, since Pashtuns on either side of the border are more likely than most others to view the Western military presence in Afghanistan as illegitimate, there is inevitably a trans-border hinterland for the insurgency. Second, since Pashtuns play a large part in the Pakistan Army, and in the Frontier Corps which comes under the Ministry of Interior, there are built-in difficulties in Pakistani government attempts to impose Islamabad's rule by force on the Pashtun-inhabited areas.[8] As a consequence, the insurgency in southern Afghanistan is likely for the foreseeable future to have safe base areas inside Pakistan. Like so many border regions in the world, the Pakistan–Afghanistan border presents excellent opportunities for the organisation and continuation of insurgency.

This leads to a third consequence of the Pakistan connection: the strong pressure on the United States to take the war unilaterally into Pakistani territory. US policy towards Pakistan notoriously lacks strategic coherence.[9] The fact that Washington considers the Pakistani authorities unreliable, with certain elements willing to pass on intelligence to America's enemies, means that the US military role on the territory of Pakistan cannot be based on close military cooperation. As a result, US military action in Pakistan is bound to be perceived as an infringement of Pakistan's sovereignty. George W. Bush's presidential order of July 2008, authorising US strikes in Pakistan without seeking the approval of the Pakistan government, while an understandable reaction to a troubling situation on the border, risks further destabilising a country that is a crucial if deeply flawed ally.[10]

Revival of doctrine

Contrary to myth, counter-insurgency campaigns can sometimes be effective. Doctrines and practices of counter-insurgency – the best of which draw on a wide and varied range of practice – have a long history.[11] The revival of counter-insurgency doctrine in the past few years has been driven primarily by events in Iraq, but also, if to a lesser degree, by the development of the insurgency in Afghanistan. This revival is hardly surprising. The response of adversaries to the extraordinary pattern of US dominance on the battlefield was always going to be one of unconventional warfare, including the methods of the guerrilla and the terrorist; and in turn the natural US counter-response was to revive the most obviously appropriate available body of military doctrine.

The key document of the US revival of counter-insurgency doctrine is US Army Field Manual 3-24.[12] It is very much an Army and Marine Corps manual: the US Air Force refused to collaborate in the exercise. Improbably for a military-doctrinal document, it has been in demand with a general readership in the United States. It has been heavily accessed on and downloaded from the Web, and is available as a published book from a major university press.[13] Although it has some flaws, it is a significant contribution to counter-insurgency literature. By contrast, the UK has not produced any major new manual. This is partly because the British had extant doctrine,[14] but also because there was some opposition to counter-insurgency doctrine on the grounds that it would result in the same hammer being used on every problem. Sir John Kiszely, until 2008 director of the Defence Academy of the UK, offered the down-to-earth reminder that 'every insurgency is *sui generis*, making generalizations problematic'.[15]

The 'comprehensive approach', central to both the US and UK doctrines, essentially means the application of all aspects of the power of the state within the territory of which the insurgency is being fought. The apparent assumption that there is a state with real power is the key weakness of the approach, especially as it applies to Afghanistan. Before exploring this in more detail, it may be useful to glance at the problematic nature of assumptions about the political realm in the counter-insurgency doctrines inherited from past eras.

The lessons of history

The US manual revives and updates doctrines that were developed in the Cold War years in response to anti-colonial insurrections (some of them involving leadership by local communist parties). It relies especially heavily on two sources.[16] The first is David Galula's *Counterinsurgency Warfare* – one of the better writings of the French thinkers on *guerre révolutionnaire*.[17] The second is Sir Robert Thompson's *Defeating Communist Insurgency*.[18] Both placed emphasis on protecting populations as distinct from killing adversaries – a crucial distinction which implies a need for high force levels.

According to the Introduction, the new US manual aspires to 'help prepare Army and Marine Corps leaders to conduct [counter-insurgency] operations anywhere in the world'.[19] This might seem to imply a universalist approach, but the authors emphasise that each insurgency is different. The foreword by Generals David Petraeus and James Amos is emphatic on this point: 'You cannot fight former Saddamists and Islamic extremists the same way you would have fought the Viet Cong, Moros, or Tupamaros; the application of principles and fundamentals to deal with each varies considerably'.[20] The US manual also emphasises the importance of constantly learning and adapting in response to the intricate environment of counter-insurgency operations – a point which strongly reflects British experience.[21]

Past exponents of counter-insurgency doctrine have generally placed heavy emphasis on achieving force ratios of about 20–25 counter-insurgents for every 1,000 residents in an area of operations. Noting this, the manual states: 'twenty counterinsurgents per 1000 residents is often considered the minimum troop density required for effective [counter-insurgency] operations; however as with any fixed ratio, such calculations remain very dependent upon the situation'.[22] This emphasis on force ratios is controversial. In any case, in Afghanistan there appears little chance of achieving such numbers. If the entire country with its 31 million inhabitants were to be viewed as the area of operations, a staggering 775,000 counter-insurgents would be needed. Even if the area of operations is defined narrowly, and even allowing for the fact that not all have to be NATO or other foreign troops, the prospects of getting close to this force ratio must be low.

There is little chance of achieving such numbers

A flaw in some past counter-insurgency doctrine has been a lack of sensitivity to context and, in some cases, an ahistorical character. Some specialists in counter-insurgency have seen their subject more as a struggle of light versus darkness than as a recurrent theme of history or an outgrowth of the problems of a society. Examples of such an ahistorical approach to the subject can be found in the French group of theorists writing in the 1950s and early 1960s about *guerre révolutionnaire*. Some denied the complexities – especially the mixture of material, moral and ideological factors – that are keys to understanding why and how guerrilla and terrorist movements come into existence. Colonel Charles Lacheroy, a leading figure in this group and head of the French Army's *Service d'Action Psychologique*, famously stated: 'in the beginning there is nothing'.[23] Terrorism was seen as having been introduced deliberately into a peaceful society by an omnipresent outside force – international communism. It is a demonological vision of a cosmic struggle in which the actual history of particular countries and ways of thinking has little or no place.

A related fault in some counter-insurgency writing was the tendency to distil general rules of counter-insurgency from particular struggles and then seek to apply them in radically different circumstances. The campaign in Malaya in the 1950s, because it was successful in ending a communist-led insurgency, was often upheld as a model, and is described favourably in the new US field manual.[24] Although they do not appear to have been elevated into formal doctrine, certain lessons drawn partly from Malaya were subsequently applied by the British in Borneo and Oman with some effect. However, successes such as that in Malaya can be great deceivers. Attempts to apply the lessons of Malaya in South Vietnam in the 1960s largely failed.[25]

The main reason was that conditions in Vietnam were utterly different from those in Malaya. In Malaya the insurgency had mainly involved the ethnic Chinese minority, and had never managed to present itself convincingly as representing the totality of the inhabitants of Malaya. The insurgency was weakened by the facts that the Chinese minority was distinguishable from other segments of society; Malaya had no common frontier with a communist state, so infiltration was difficult; and the British granting of independence to Malaya undermined the anti-colonial credentials of the insurgents. In South Vietnam, by contrast, the communist insurgents had strong nationalist credentials, having fought for independence rather than merely having power handed to them by a departing colonial power.[26] At the heart of the US tragedy in Vietnam was a failure to recognise the unique circumstance of the case – that in Vietnam, more than any other country in Southeast Asia, communism and nationalism were inextricably intertwined.

One lesson that could have been drawn from the Malayan case is that it is sometimes necessary to withdraw to win. The new US manual places much emphasis on the fact that the United States withdrew from Vietnam in 1973 only to see Saigon fall to North Vietnamese forces in 1975.[27] It does not note a contrary case: the UK promise to withdraw completely – a promise that was followed by the Federation of Malaya's independence in 1957 – contributed to the defeat of the insurgency in Malaya.[28] The value of such promises needs to be taken into account in contemporary counter-insurgency efforts and indeed counter-insurgency theory. This is especially so, as the idea that the United States intended to stay indefinitely in Iraq and Afghanistan, and to build networks of bases there, had a corrosive effect in both countries and more generally. The decision of the Iraqi cabinet on 16 November 2008 that all US forces will withdraw from Iraq by 2011 is evidence that a guarantee of withdrawal is seen as a necessary condition (and not simply a natural consequence) of ending an acute phase of insurgency.

Other weaknesses

One weakness in the US manual, likely to be remedied in any future revisions, is the lack of serious coverage of systems of justice – especially those employed by the insurgents themselves. The references to judicial systems in the manual are brief and anodyne, almost entirely ignoring the challenge posed by insurgents in this area.[29] Insurgencies commonly use their own judicial procedures to reinforce their claims to be able to preserve an existing social order or create a better one. The Taliban have always placed emphasis on provision of a system of Islamic justice.[30] In the current conflict, taking advantage of the fact that the governmental legal system is weak and corrupt, they have done this effectively in parts of Afghanistan.

This leads to a more general criticism. In addressing the problem of undermining and weakening insurgencies, both traditional counter-insurgency theory and its revived versions in the twenty-first century place emphasis on the role of state institutions: political structures, the administrative bureaucracy, the police, the courts and the armed forces. The institutions are often taken for granted, and assumed to be strong. Indeed, the current British counter-insurgency doctrine stemmed from a project started in 1995 to capture the lessons and doctrine from Northern Ireland. A common criticism of much counter-insurgency practice is that it was enthusiastically pursued by over-powerful and thuggish states, especially in Latin America.[31]

Today, counter-insurgency theories risk being out of joint with the realities of assisting the so-called 'failed states' and 'transitional administrations' of the twenty-first century. These problems are not new; one of the problems that undermined US counter-insurgency efforts in Vietnam was the artificiality and weakness of the coup-prone state of South Vietnam. Yet the central fact must be faced that in the two test-beds of the new counter-insurgency doctrines of recent years, Iraq and Afghanistan, state institutions have been notoriously weak. Indeed, in post-colonial states generally, where insurgencies are by no means uncommon, indigenous state systems tend to be fragile or contested. The role of the state in people's lives, and in their consciousness, may be thoroughly peripheral or even negative.[32] So when the US manual speaks of 'a comprehensive strategy employing all instruments of national power' and stresses that all efforts focus on 'supporting the local populace and [host-nation] government',[33] it is necessary to remind ourselves that support for government is not exactly a natural

State institutions have been notoriously weak

default position for inhabitants of countries with such tragic histories as Iraq and Afghanistan. On the other hand, General Petraeus worked on the manual after two tours of duty in Iraq, with an eye to applying it there, and then did so to some effect when he was commander of the Multinational Force-Iraq. In 2009 the Iraqi government is looking stronger than in the first years after the invasion. The fact that a government is weak in face of insurgency does not mean that it is necessarily fated to remain so.

Of the many critiques of the US revival of counter-insurgency doctrine, one of the most searching is an American Political Science Association review symposium published in June 2008.[34] Stephen Biddle of the US Council on Foreign Relations queried the manual's fundamental assumption when he stated that

> it is far from clear that the manual's central prescription of drying up an insurgent's support base by persuading an uncommitted population to side with the government makes much sense in an identity war where the government's ethnic or sectarian identification means that it will be seen as an existential threat to the security of rival internal groups, and where there may be little or no supracommunal, national identity to counterpose to the subnational identities over which the war is waged by the time the United States becomes involved.[35]

Biddle also pointed out that the US manual has little to say about the comparative merits of waging counter-insurgency with large conventional forces as against small commando detachments, on the relative utility of air power in counter-insurgency, and on the willingness of democracies to support counter-insurgency over a long period. Further, the manual does not fit particularly well the realities of Iraq, where the insurgencies are far more regional and localised in character, and more fickle in their loyalties, than were many of the communist and anti-colonial insurgencies of earlier eras. As Biddle says, the negotiation of local ceasefires between insurgents and US commanders has been of key importance in Iraq.[36] Such webs of local ceasefires, valuable despite their fragility, do not come from counter-insurgency doctrine. These criticisms are another way of saying what General Petraeus knows, that all doctrine is interim, and some parts are more interim than others.

The need to adapt doctrine, so evident in Iraq, applies even more strongly to Afghanistan, a subject about which the US manual says remarkably little.[37] The key issue is whether the revival of counter-insurgency doctrine really offers a useful guide in a situation where there are some distinct elements in the insurgencies, where negotiation with some of the insurgents may have a role, and where the state does not command the same loyalty or obedience that more local forces may enjoy.

After a difficult year in 2008, the US and Afghan governments began to place increased emphasis on local social structures. The US Ambassador to Afghanistan said at the end of the year that there was agreement to move forward with two programmes: first, the community outreach programme, 'designed to create community shuras' (local councils); and second, the community guard programme, which is 'meant to strengthen local communities and local tribes in their ability to protect what they consider to be their traditional homes'.[38] While neither programme was well defined, the move in this direction was evidence of willingness to rely on a less state-based approach than hitherto.

Air power

Since October 2001 air power, which mainly means US air power, has played an important part in military operations in Afghanistan. The apparent success of the use of air power in October–December 2001 was deceptive: a major factor in the Taliban's defeat was the advance of ground forces of the Northern Alliance. Since then, the role of air power in the Afghan conflict has been a subject of contestation, principally between the army and marines on the one hand, and the US Air Force on the other. A key issue has been whether air power is a major instrument in its own right, or is mainly useful in supporting ground forces. Self-evidently, the US and NATO ground forces in Afghanistan, widely dispersed and few in number, frequently need air power in support of their ground operations. In military terms, a 'light footprint' on the ground inevitably means a heavy air presence.

Those planning coalition military operations in Afghanistan have shown awareness of the dangers of reliance on air power – especially of the adverse consequences of killing civilians. On occasion they have even claimed to have set an aim of no civilian casualties.[39] While this aim actually goes further than the strict requirements of existing law applicable in an international armed conflict, in practice it has not been achieved. Many factors have prevented its realisation: shortage of ground forces, differing approaches of individual commanders, poor intelligence, the heat of battle, weapons malfunction, the co-location of military targets and civilians, and the frayed relationship between ground and air forces operating in Afghanistan.[40] A Human Rights Watch report in September 2008 summarised the situation:

> In the past three years, the armed conflict in Afghanistan has intensified, with daily fighting between the Taliban and other anti-government insurgents against Afghan government forces and its international military supporters. The US, which operates in Afghanistan through its counter-insurgency forces in Operation Enduring Freedom (OEF) and as part of the NATO-led International Security Assistance Force (ISAF), has increasingly relied on airpower in counter-insurgency and counter-terrorism operations. The combination of light ground forces and overwhelming airpower has become the dominant doctrine of war for the US in Afghanistan. The result has been large numbers of civilian casualties, controversy over the continued use of airpower in Afghanistan, and intense criticism of US and NATO forces by Afghan political leaders and the general public.

As a result of *Operation Enduring Freedom* and ISAF airstrikes in 2006, 116 Afghan civilians were killed in 13 bombings. In 2007, Afghan civilian deaths were nearly three times higher: 321 Afghan civilians were killed in 22 bombings, while hundreds more were injured. In 2007, more Afghan civilians were killed by airstrikes than by US and NATO ground fire. In the first seven months of 2008, the latest period for which data are available, at least 119 Afghan civilians were killed in 12 airstrikes.[41] That last figure increased dramatically when it was revealed in October 2008 that 33 civilians had been killed in a single US airstrike on 22 August. Such incidents do serious damage to the coalition cause. Largely as a result of the long history of such incidents, there has been a strong anti-coalition reaction. As early as 2006 the Afghan parliament had demonstrated its concern about coalition military actions, and such expressions of concern have subsequently become more frequent. Meanwhile, President Hamid Karzai, whose authority has been diminishing, has made a number of criticisms of the coalition forces, calling for an end to civilian casualties, and even stating that he wanted US forces to stop arresting suspected Taliban members and their supporters.[42]

Judging progress

Assessing results in counter-insurgency wars is by nature a contentious task, and involves difficult questions about the appropriate methodologies. Sometimes unorthodox methods of analysis yield the most valuable answers. The war in French Indochina from 1946 to 1954 provided a classic case. When a French doctoral student, Bernard Fall, went to Vietnam in 1953, the French authorities claimed that the war was going well, and showed maps and statistics indicating that they controlled a large proportion of the territory. But Fall soon realised that French claims about the amount of territory they controlled were exaggerated, or at least lacked real meaning as far as the conduct of government was concerned. He reached this conclusion both by visiting Vietminh-held areas, and by inspecting tax records in supposedly government-held areas: these latter showed a dramatic collapse in the payment of taxes, and thus indicated a lack of actual government control.[43] In Afghanistan, payment of taxes, or rather the absence of payment, works as a measure of the government's lack of control. As Astri Suhrke has shown, taxation constitutes a uniquely small proportion – in 2005 it was only 8% – of all estimated income in the national budget.[44]

By one key measure serious progress may appear to be being made in the Afghan war. The numbers of refugees returning to Afghanistan since the fall of the Taliban regime at the end of 2001 are one possible indicator of a degree of progress. According to The United Nations High Commissioner for Refugees (UNHCR), which played a key part in the process, between 1 January 2002 and 31 December 2007 4,997,455 refugees returned to

Table 1. **Refugees returning to Afghanistan, 2002–2007**

2002	2003	2004	2005	2006	2007
1,957,958	645,864	879,780	752,084	387,917	373,852

Afghanistan (see Table 1). This is the largest refugee return in the world in a generation. It is striking that even in 2006 and 2007 (years of considerable conflict in parts of Afghanistan) the returns continued, if at a reduced rate. In the whole period 2002–07, the overwhelming majority of refugees have been in two countries: Iran, from which 1.6m returned, and Pakistan, from which 3.3m returned.[45] Impressive as these figures are, four major qualifications have to be made:

- First, they have to be understood against the backdrop of the sheer numbers of Afghan refugees: at the end of 2007 Afghanistan was still the leading country of origin of refugees worldwide, with 3.1m remaining outside the country. Thus in 2008, even after these returns, Afghan refugees constitute 27% of the entire global refugee population.
- Secondly, not all returns were fully voluntary. Within the countries of asylum there have been heavy pressures on these refugees to return, including the closing of some camps.
- Thirdly, experience of many returning refugees has included lack of employment opportunities in Afghanistan, and in some cases involvement in property disputes. There has been mismanagement and corruption in the Afghan Ministry of Refugees and Repatriation. Some refugees live in dire conditions in makeshift settlements. All this has created much disappointment, bitterness and anti-government feeling.
- Fourthly, displacement continues. In the past two years unknown numbers of returnees have left the country again. The number of internally displaced persons within Afghanistan has also increased, especially due to the fighting in the south, and now stands at about 235,000. Some returnees have seamlessly become internally displaced persons.[46]

Other developments confirm this sobering picture. The Afghan army remains relatively small and highly dependent on outside support. As for the insurgent forces, they appear to have no shortage of recruits. Large numbers of fighters are able to cross into Afghanistan, mainly from Pakistan, and the Taliban can also employ many locals, especially in seasons when other work is in short supply. The fact that the estimated unemployment rate is 40% means that insurgents continue to have opportunities for recruitment. In Kabul and other cities, terrorist attacks, once rare, have become common. Serious observers have reported an atmosphere of disappointment and bitterness in Afghanistan in 2008.[47]

Insurgent forces appear to have no shortage of recruits

The UN Secretary-General's report of September 2008 states bluntly that the number of security incidents rose to 983 in August 2008, the highest since the fall of the Taliban in 2001 and a 44% increase year-on-year, and that 'insurgent influence has intensified in areas that were previously relatively calm, including in the provinces closest to Kabul'.[48] It reports some successes in the campaign against poppy cultivation, and it strongly endorses the Afghanistan National Development Strategy, adopted at the Paris Conference in Support of Afghanistan, held on 12 June 2008. However, it confirms the picture of the state of progress in the war against the Taliban which has also been depicted by other sources, including the sober report by General David McKiernan, the top US Commander in Afghanistan, who, at the same time as seeking specific troop increases, has rejected simple notions, and indeed the terminology, of a military 'surge';[49] and the US National Intelligence Estimate on Afghanistan, a draft version of which was leaked in October 2008, and which stated that the situation there was in a 'downward spiral'.[50] One grim statistic of the downward spiral is the casualty rate of IFOR and *Operation Enduring Freedom* forces in Afghanistan. Fatalities have increased each year from 57 in 2003 to 296 in 2008.[51]

As so often in counter-insurgency wars, the most useful assessments may be those of independent witnesses who, just as Bernard Fall did in French Indochina, have deep knowledge of a society and a healthy open-mindedness about the contribution that outside forces can make to security. Rory Stewart, who walked across Afghanistan in 2002, and later retired from the UK diplomatic service to run a charitable foundation in Kabul, is perhaps Fall's nearest equivalent today. He has argued that 'we need less investment – but a greater focus on what we know how to do'. He is specifically critical of increases in forces:

> a troop increase is likely to inflame Afghan nationalism because Afghans are more anti-foreign than we acknowledge and the support for our presence in the insurgency areas is declining. The Taliban, which was a largely discredited and backward movement, gains support by portraying itself as fighting for Islam and Afghanistan against a foreign military occupation.[52]

The present campaign in Afghanistan is unlikely to result in a clear victory for the Kabul government and its outside partners, because the sources of division within and around Afghanistan are just too deep, and the tendency to react against the presence of foreign forces too ingrained. The war could yet be lost, or, perhaps more likely, it could produce a stalemate or a long war of attrition with no clear outcome. The dissolution of Afghanistan into regional fiefdoms – already an accustomed part of life – could continue and even accelerate.

To some it may appear remarkable that Afghanistan has not reverted more completely to type as a society that rejects outside intrusion. Part of the explanation may be that this is not the only natural 'default position' for Afghans: there have also been countless episodes in which Afghan leaders have sought, and profited from, alliances with outsiders. A second factor is the 'light footprint' advocated by Lakhdar Brahimi, special representative of the secretary-general for Afghanistan: for all the limitations of this approach, and the many departures from it since it was enunciated in 2002 with specific reference to the UN Assistance Mission in Afghanistan (UNAMA), no one has convincingly suggested a better one.[53] A third factor is that – notwithstanding the disastrous killings of civilians as a result of using air power – there has been a degree of restraint in the use of armed force: this has been important in at least slowing the pace of the process whereby the United States and other outside forces come to be perceived as alien bodies in Afghanistan. The interesting phenomenon of application of certain parts of the law of armed conflict – namely the rules of targeting – as if this was an international war is part of this process.

The role of the United Nations

The UN has some remarkable achievements to its credit in Afghanistan. It helped to negotiate the Soviet withdrawal, and since then it has remained engaged on the ground. It gave a degree of authorisation to the US-led effort to remove the Taliban regime in 2001, it has authorised ISAF and has provided a legitimate basis for its expanded roles throughout the country, and it has been involved in the many subsequent efforts to help develop Afghanistan, not least by assisting in the various elections held there since 2001. It has assisted the largest refugee return to any country since the 1970s.

Despite these achievements, the UN's roles have been more limited than those of the United States and its partners, especially in matters relating to security. That the UN's role in this crisis has been modest is not especially surprising. Neither the terms of the UN Charter nor the record of the Security Council justify the excessively high expectations many have had in respect of the council's roles. It was always a mistake to view the UN as aiming to provide a complete system of collective security even in the best of circumstances – and circumstances in and around Afghanistan are far from being favourable for international involvement.

International legitimacy, moreover, is never a substitute for local legitimacy. The council's acceptance of regime change in Afghanistan was justified once the Taliban had refused to remove al-Qaeda, and did much to legitimise the aim of regime replacement, which could otherwise have seemed a narrowly neo-colonial US action. Yet there is a danger that such international conferrals of legitimacy can contribute to a failure to address the no less important question of securing legitimacy in the eyes of the audience that matters most: in this case, the peoples of Afghanistan and neighbouring countries.

The NATO framework

In Afghanistan NATO is involved in ground-combat operations for the first time in its history, far from its normal area of responsibility and against a threat very different from the one it had been created to face. NATO involvement in Afghanistan is widely viewed as 'a test of the alliance's political will and military capabilities'.[54] It is an exceptionally hard test. Indeed, the implication that the future of the Alliance hangs on this test is reminiscent of earlier views that US credibility was on the line in Vietnam.

NATO's involvement in Afghanistan is in sharp contrast to its conduct during the Cold War, when it repeatedly and studiously avoided involvement in colonial conflicts – the French wars in Indochina and Algeria, the Portuguese wars in Africa, the British in Malaya, the Dutch in Indonesia and so on. Individual members were involved, but the alliance was not. NATO also avoided involvement in post-colonial conflicts or, as in Cyprus, limited itself to an essentially diplomatic role. Now NATO has become engaged, with little public debate, in a country with all the hallmarks of a post-colonial state undergoing conflict, especially the lack of legitimacy of the constitutional system, government and frontiers.

NATO's role in Afghanistan began in a problematic way, and so it has continued. The day after the 11 September attack, the NATO Council stated: 'if it is determined that this attack was directed from abroad against the United States, it shall be regarded as an action covered by Article 5 of the Washington Treaty, which states that an armed attack against one or more of the Allies in Europe or North America shall be considered an attack against them all'.[55] When Washington gave this offer the brush-off, preferring to have a 'coalition à la carte' in which there would be no institutional challenge to its leadership, there was disappointment and irritation in Europe. The war in Afghanistan in October–December 2001, while effectively conducted under US leadership, was also one chapter in the story of the declining size of US-led wartime coalitions.

However, NATO rapidly came back into the picture, not least because Washington came to recognise the need for long-term assistance in managing societies that had been freed from oppressive regimes by US force. NATO has been directly involved in Afghanistan at least since 9 August

2003, when it took formal control of ISAF. It was in autumn 2003 that an upsurge of violence that was part of a deteriorating security situation began.[56] ISAF's notably broad UN Security Council mandate involves it in a wide range of activities, including reconstruction and economic development, and military and police training. Since 2006 ISAF has undertaken an expanded range of responsibilities in Afghanistan, involving combat as well as peacekeeping, in an expanded area which includes provinces in which conflict is ongoing.

It is truly remarkable that the reputation of the longest-lived military alliance in the world, comprised of states with fundamentally stable political systems, should have made itself vulnerable to the outcome of a war in the unpromising surroundings of Afghanistan. There is much nervousness about this among NATO's European members, and this may explain the reluctance of European leaders to make the kind of ringing statements that often accompany war. Knowing that the outcome of any adventure in Afghanistan is bound to be uncertain, they have wisely kept the level of rhetoric low.

European leaders have wisely kept the level of rhetoric low

There may be another reason for the reluctance of many leaders of European member states to make strong endorsements of their participation in the war in Afghanistan. Many of the claims that can be made in favour of the Afghan cause are also implicitly criticisms of the involvement in Iraq. From the start in 2001, the US-led involvement in Afghanistan and the subsequent involvement of ISAF have both had a strong basis of international legitimacy that was reflected in Security Council resolutions. In Afghanistan there was a real political and military force to support, in the shape of the Northern Alliance. In Afghanistan and Pakistan there were real havens for terrorists. In Afghanistan, up to 5m refugees have returned since 2001. To speak about these matters too loudly might be to undermine the US position in Iraq, where the origins and course of the outside involvement have been different, and where the flow of refugees has been outwards. NATO leaders, anxious to put the recriminations of 2003 over Iraq behind them, may be nervous about highlighting the differences between Afghanistan and Iraq.

A major question, heavy with implications for international security, is how the setbacks experienced in Afghanistan are to be explained, especially within NATO member states. The UN may be accustomed to failure, but NATO is not. So far, the tendency has been to blame Pakistan, the messy NATO command, the poor attention span of US governments, the unwillingness of NATO allies to contribute, the weakness of Karzai, the corruption of his government, the shortage of foreign money and troops – in other words, to blame almost everything except the nature of the project.

The various reasons that have been given cannot be lightly dismissed. For example, the lack of NATO unity in certain operational matters has been striking. Different contributing states have different visions of ISAF's role. The most obvious difference is that the United States, Britain and Canada tend to see it, albeit with some variations within each country, as encompassing a counter-insurgency operation, while Germany and some others see it more through the lens of a stabilisation mission. These positions are not polar opposites, and each may have validity in different provinces of Afghanistan, but the clash of perspective on this issue does not assist cooperation of forces in difficult operations. Daniel Marston has gone so far as to conclude that, 'as of 2007, the main problem impeding coalition forces' successful application of counter-insurgency was decentralization of responsibility'.[57] The inability of member states to agree on a straightforward and defensible common set of standards for treating prisoners in the Afghan operations is symptomatic of deep divisions within the alliance. Anxious not to be associated with shocking US statements and practices in this matter, and insufficiently staffed and equipped to hold on to the prisoners they capture, other NATO members have drawn up separate agreements with the Afghan authorities, embodying a variety of different approaches to how prisoners should be treated once in Afghan hands. There are serious concerns that some detainees handed over to the Afghan authorities on this basis have been maltreated.[58] The complexity of the command-and-control arrangements in Afghanistan is greater than that in past counter-insurgency campaigns. Debates about this have inevitably reflected the US desire that more contingents in ISAF should become directly involved in combat operations, and the concern of some contributors that this should not happen.

Although ISAF is now under a US commander, and the continuous rotation of senior posts is ceasing, the arrangements for coordinating the work of the three distinct forces – ISAF, the Afghan National Army and the US-led *Operation Enduring Freedom* – continue to pose problems.[59]

The provision of forces in the numbers required for ISAF has been a highly contentious matter within NATO states. Granted the scale of the problems in Afghanistan, the levels of the three distinct forces are widely seen as low, yet in many NATO member states there is a reluctance to increase the commitment. Opinion polls in five NATO member states with a high level of involvement in Afghanistan show the public to be highly sceptical about it.[60] An increase in such numbers risks running into opposition in many NATO states, and also further antagonising Afghan opinion.

Political divisions have never been far from the surface, and will no doubt be projected into future explanations of what went wrong. Continental Europeans can convincingly blame the Americans and the British for having taken their eye off the ball in Afghanistan in 2002–03, foolishly thinking that the war there was virtually won and that they could afford to rush into a second adventure in Iraq. Americans can blame the Europeans for putting relatively few troops into ISAF, and being slow to back them up when the going got rough in 2006–08. A less blame-centred explanation might be that the reconstruction of Afghanistan, and the pursuit of counter-insurgency there, was always going to be an extremely difficult task; that there are limits to what outsiders should expect to achieve in the transformation of distant societies with cultures significantly different from their own; and that it never made sense to invest such effort in counter-insurgency in Afghanistan without having even the beginnings of a strategy for the neighbouring regions of Pakistan.

Impact on international security

The problem of Afghanistan – including the complex interplay of international actors who have pursued their interests there – has had an impressive impact on international security issues in the past generation. It contributed to the end of the Cold War and indeed of the Soviet Union itself. The Taliban regime's failure to control al-Qaeda activities launched the United States into the huge and seemingly endless 'war on terror', and also resulted in the UN Security Council claiming unprecedented powers to affect activities within states. The Afghan war has embroiled NATO in a largely civil war thousands of miles from its North Atlantic heartlands.

For the future, the greatest impact of Afghanistan on international security may turn out to be highly paradoxical. It is obvious that Afghanistan, along with Iraq, has called into question the idea that the United States, in its supposed 'unipolar moment', could change even the most difficult and divided societies by its confident use of armed force. But it is not only the ideas of the neo-conservatives and their camp-followers that are in trouble. In many ways the involvement of NATO in Afghanistan was textbook liberal multilateralism: approved by the UN Security Council, involving troops from 40 democracies, coupled with the UN Assistance Mission, and with admirable aims to assist the development and modernisation of Afghanistan. The very ideas of rebuilding the world in the Western image, and of major Western states having an obligation to achieve these tasks in distant lands – whether by unilateral or multilateral approaches – may come to be viewed as optimistic. Or, to put it differently, and somewhat cryptically, Afghanistan may not have quite such a drastic effect on the American *imperium* as it had on the Soviet one in the years up to 1991; but it may nevertheless come to be seen as one important stage on the path in which international order became, certainly not unipolar, and perhaps not even multipolar, but more based on prudent interest than on illusions that Western ideas control the world. Afghanistan may contribute to greater caution before engaging in interventionist projects aimed at reconstructing divided societies.

NATO in Afghanistan was textbook liberal multilateralism

Despite all the difficulties encountered in Afghanistan since the fall of the Taliban in 2001, in the US presidential election campaign in 2008 both Barack Obama and John McCain promised to increase the US commitment to Afghanistan in 2009. There was little prospect either that the insurgency would subside or that Washington would tip-toe out of the war. Furthermore, both candidates advocated continuing and even extending the practice of using US force against Taliban and al-Qaeda targets in Pakistan. The war's international dimension, and its significance for international security more generally, was set to continue.

* * *

The Obama administration's policy planning for Afghanistan is based on the sound presumption that the Afghan problem cannot be addressed in isolation. Although many countries have a potentially important role in any settlement in Afghanistan – especially Iran, with its large numbers of Afghan refugees and its major drug problem – Pakistan is at the core of this approach. Granted the indissoluble connection between Afghanistan and Pakistan, any policy in respect of the one has to be framed in light of its effects on the other. At times it may even be necessary to prioritise as between these two countries. The simple truth is that Pakistan is a far larger, more powerful and generally more important country than Afghanistan. If the price of saving Afghanistan were to be the destabilisation of Pakistan, it would not be worth paying. A principal aim of the United States in the region should have been, and indeed may have been, to avoid creating a situation in which that particular price had to be paid: yet at least once before, in the Soviet–Afghan War in the 1980s, that, or something very like it, is what happened.

The main conclusion of any consideration of the Pakistan factor in the ongoing conflict in Afghanistan has to be that the policy of the United States and allies – to strengthen central government in both countries – has been operating in extremely difficult circumstances, has been pursued erratically, and has been largely unsuccessful. While it is not obvious what the alternatives might be, the general approach of backing non-Pashtuns in Pakistan and Afghanistan risks exacerbating the Pashtun problem in both countries. Three distinct causes – Pashtun, Taliban and al-Qaeda – have become conflated. It should be a first aim of Western policy to reverse this dangerous trend.

Because of the grim prospects of a stalemate, a war of attrition or worse in Afghanistan, and also because of the advent of new governments in Pakistan in 2008 and the US in 2009, there has been at least the beginning of consideration of alternative policies. Two stand out: each in its way addresses directly the growth of the insurgency, and is based on a recognition that the Pakistan dimension of the problem has to be considered alongside the Afghan. Both options take into account the central requirement of any approach, that it be geared to ensuring that neither Afghanistan nor Pakistan offer the kind of haven for organising international terrorist actions that Afghanistan did under Taliban rule.

The first option centres on negotiation with Taliban and other Pashtun groups. The first question to be faced is whether, on either side of the border, there are sufficiently clear hierarchical organisational structures with which to negotiate. The second question is whether Afghan Taliban/Pashtun goals are framed more in terms of control of the Afghan state along the completely uncompromising lines followed by the Taliban in the years up to 2001, or in more limited terms. Whatever the answers, negotiation in some form with some of the insurgent groups and factions is inevitable. Indeed, in an informal manner some is already happening. Combining fighting with talking is quite common in insurgencies, not least because of their tendency to result in stalemate. Yet it is never easy, and is likely to be particularly difficult for those on both sides who have chosen to see the war in Afghanistan as a war of good against evil. It is also likely to be difficult if, as at present, the Taliban believe they are in a position of strength. A critical question to be explored in any talks is whether, as some evidence suggests, the Taliban leaders have learned enough from their disasters since seizing Kabul in 1996, and in particular from their near-death experience in 2001, to be willing to operate in a different manner in today's Afghanistan.[61] The scope and content of any agreement are matters of huge difficulty. Some agreements concluded by the Pakistan government in the past few years are widely seen as having given Taliban leaders a licence to continue supporting the insurgency in Afghanistan. This serves as a warning of the hazards of partial negotiation. Yet the pressures for negotiation are very strong, and a refusal to consider this course could have adverse effects in both countries.

In October 2008, after a two-week debate that was not always well attended, the Pakistani parliament passed unanimously a resolution widely interpreted as suggesting, above all, a shift to negotiation. It was a complex package, in which the parliament united to condemn terrorism and at the same time was seen as 'taking ownership' of policy to tackle it. The resolution said that regions on the Afghan border where militants flourish should be developed, and force used as a last resort. It opposed the cross-border strikes by US forces in Pakistan, but at the same time indicated a degree of support for US policy. It called for dialogue with extremist groups operating in the country, and hinted at a fundamental change in Pakistan's approach to the problem: 'we need an urgent review of our national security strategy and revisiting the methodology of combating terrorism in order to restore peace and stability'.[62] At the very least it provides one basis for the incoming US administration to recalibrate the United States' largely burnt-out policies towards Pakistan.

The second option under discussion involves a fundamental rethinking of security strategy in both Afghanistan and Pakistan. On the Afghan side of the border it would call for some increase in ISAF or other outside forces, especially to speed up the pace of expansion of the Afghan army, and thereby provide back-up so that areas from which the Taliban have been expelled can thereafter be protected. It would also call for cooperation in security matters with local forces and councils, with all the hazards involved; for expansion of aid and development programmes, especially in urgent matters such as food aid in areas threatened by famine; and for a serious effort to address the widespread corruption which makes a continuous mockery of Western attempts to bring reform and progress to Afghanistan. On the Pakistan side it would involve a protracted effort to develop a long-term policy (hitherto non-existent) for establishing some kind of government influence in the Federally Administered Tribal Areas, and for a joined-up policy for addressing the Taliban and al-Qaeda presence. On both sides of the border it would necessitate reining in the use of air power in ways that reduce its inflammation of local opinion.

It is highly improbable that either of these options on its own could provide a substantial amelioration of a tangled and tragic situation. However, a combination of the two – both negotiating, and rethinking the security strategy – might just achieve some results.[63] It would need to include other elements, including a strong and credible commitment to leave as soon as a modicum of stability is achieved. Such a combination would need to be pursued in both Afghanistan and Pakistan. It could only work if the new US administration rejected the worst aspects of previous policies, and pursued the matter with more consistent attention than in the past. It would be likely to result in some unsatisfactory compromises, and might build on, rather than fundamentally change, the pattern of regional warlordism so rooted in Afghanistan. Yet if the war in Afghanistan is not to have even more fateful consequences for international order than those seen in the past three decades, it may be the direction in which events have to move.

Acknowledgements

This essay is a product of research conducted under the auspices of the Oxford University Leverhulme Programme on the Changing Character of War. It is based on a presentation at the US Naval War College International Law Department experts' workshop on 'The War in Afghanistan', 25–27 June 2008. A fuller version of the essay will appear in due course in Michael N. Schmitt (ed.), *The War in Afghanistan: A Legal Analysis* (Newport, RI: US Naval War College International Law Studies vol. 85). I am grateful to Alex Alderson, Jeremy Allouche, John Nagl, Hew Strachan, Astri Suhrke and Susan Woodward for their comments on earlier drafts of this paper. The usual disclaimer, that all responsibility for error is mine and mine alone, applies with particular force in this case.

Notes

1 On the revolution in military affairs and related doctrines, and how their weaknesses became evident, see Lawrence Freedman, *The Transformation of Strategic Affairs*, Adelphi Paper 379 (Abingdon: Routledge for the International Institute for Strategic Studies, London, 2006).

2 Paul Wolfowitz, on CBS TV, 'Face the Nation', 18 November 2001, transcript available at http://www.defenselink.mil/transcripts/transcript.aspx?transcriptid=2442.

3 David Loyn, *Butcher and Bolt: Two Hundred Years of Foreign Engagement in Afghanistan* (London: Hutchinson, 2008).

4 On the Taliban's history of supporting opium production, which became the mainstay of their war economy in the late 1990s, see Ahmed Rashid, *Taliban: The Story of the Afghan Warlords* (London: Pan Books, 2001), pp. 117–24.

5 Ahmed Rashid, *Descent into Chaos: How the War against Islamic Extremism is Being Lost in Pakistan, Afghanistan and Central Asia* (London: Allen Lane, 2008), p. 401.

6 UN Security Council Resolution 1510 of 13 October 2003 expanded

ISAF's sphere of operations to other parts of Afghanistan. By the end of 2006, now operating under NATO, it had responsibilities in virtually all Afghanistan.

7 Rashid, *Taliban*, esp. pp. 26–9, 45, 90–94 and 137–8.

8 On the extent of Pakistani help to the Taliban, see Rashid, *Descent into Chaos*; and Seth Jones, *Counterinsurgency in Afghanistan* (California: RAND Corporation, June 2008), available at http://www.rand.org/hot_topics/afghanistan.html.

9 For an indictment predicated on the assumption that a serious policy could be devised see US Government Accountability Office, *Combating Terrorism: The United States Lacks Comprehensive Plan to Destroy the Terrorist Threat and Close the Safe Haven in Pakistan's Federally Administered Tribal Areas* (Washington DC: GAO, April 2008).

10 Eric Schmitt and Mark Mazzetti, 'Bush Said to Give Orders Allowing Raids in Pakistan', *New York Times*, 10 September 2008.

11 For an excellent overview, from the late nineteenth century to Afghanistan, see Marston and Malkasian (eds), *Counterinsurgency in Modern Warfare* (Oxford: Osprey, 2008). Marston's chapter is notably critical of the failure of the United States and its allies to train and equip soldiers for counter-insurgency (p. 220).

12 *Counterinsurgency*, US Army Field Manual 3-24 and Marine Corps Warfighting Publication 3-33.5 (Washington, DC: Department of the Army et al., 15 December 2006). This publication has a short foreword by Lt-Gen. David H. Petraeus (who played a key part in its preparation) and Lt-Gen. James F. Amos. Henceforth, *US Army Field Manual 3-24*. Available at http://www.fas.org/irp/doddir/army/fm3-24.pdf.

13 The manual was issued by a publishing firm as *The US Army / Marine Corps Counterinsurgency Field Manual: US Army Field Manual no. 3-24 and Marine Corps Warfighting Publication no. 3-33.5* (Chicago, IL: University of Chicago Press, 2007). This edition has a new foreword by Lt-Col. John A. Nagl, and a new Introduction by Sarah Sewall.

14 UK Army Field Manual, vol. 1, *The Fundamentals*, Part 10, *Counter Insurgency Operations (Strategic and Operational Guidelines)*, last revised in July 2001. The approach it laid out and its principles are still regarded as valid. Its biggest problem was the context in which it was set. It makes no mention of coalition operations, or the problems of operating in other people's countries, the religious and cultural dimensions, and the effects of information proliferation and information operations. The task of updating it started in late 2005. It is still in development. The Ministry of Defence's Joint Doctrine & Concepts Centre's short (23 pages) *The Comprehensive Approach* is a more general survey intended to be relevant to a wide range of operations; the word 'counter-insurgency' does not appear in it. *The Comprehensive Approach*, Joint Discussion Note 4/05 (Shrivenham, Wiltshire: JDCC, January 2006), available at http://www.mod.uk/DefenceInternet/MicroSite/DCDC/

OurPublications/JDNP/. The UK Ministry of Defence Joint Doctrine Note 2/07, *Countering Irregular Activity Within a Comprehensive Approach* (UK Ministry of Defence, March 2007), has not gone into general public circulation and has not been greeted with enthusiasm in the army.

15 John Kiszely, *Post-modern Challenges for Modern Warriors*, Shrivenham Papers No. 5 (Shrivenham: Defence Academy of the UK, 2007), p. 14.

16 *US Army Field Manual 3-24*, Acknowledgements, p. viii. Three sources, all cited at length in the text, are listed at this point. (The third, not discussed here, was an article in the *New Yorker* in January 2005.) See also the Annotated Bibliography at the end, which cites a wider range of sources. It omits key critical writings on the subject, most notably Peter Paret, *French Revolutionary Warfare from Indochina to Algeria: The Analysis of a Political and Military Doctrine* (London: Pall Mall Press, 1964). The omission of this title reflected a view that it is hard to get Americans to take on board French doctrines on counter-insurgency.

17 David Galula, *Counterinsurgency Warfare: Theory and Practice* (London: Pall Mall Press, 1964). Galula died in 1968. His work was belatedly published in France as *Contre-Insurrection: Théorie et pratique* (Paris: Economica, 2008).

18 Sir Robert Thompson, *Defeating Communist Insurgency: Experiences from Malaya and Vietnam* (London: Chatto & Windus, 1966).

19 *US Army Field Manual 3-24*, Introduction, p. ix.

20 *US Army Field Manual 3-24*, Foreword. The Moros, perhaps the least known of the insurgents cited, have been involved in an armed insurrection in the Philippines.

21 *US Army Field Manual 3-24*, p. 5-31.

22 *US Army Field Manual 3-24*, p. 1-13.

23 Col. Charles Lacheroy, 'La Guerre Révolutionnaire', talk on 2 July 1957 reprinted in *La Défense Nationale*, Paris, 1958, p. 322; cited in Paret, *French Revolutionary Warfare from Indochina to Algeria* (London: Pall Mall Press, 1964), p. 15. Paret comments that 'nothing', in this case, means 'the secure existence of the *status quo*'.

24 *US Army Field Manual 3-24*, pp. 6-21 and 6-22.

25 See especially Thompson, *Defeating Communist Insurgency*, pp. 17–20.

26 The geographical, sociological, political and ethnic differences between Malaya and South Vietnam were evident to knowledgeable observers even while the Vietnam War was still ongoing. See Bernard B. Fall, *The Two Viet-Nams: A Political and Military Analysis* (London: Pall Mall Press, 1963), pp. 339–40 and 372–6.

27 *US Army Field Manual 3-24*, pp. 1-8 and 2-13.

28 See, for example, the obituary of Sir Donald MacGillivray, the last British High Commissioner for Malaya, *Times*, 28 December 1966.

29 *US Army Field Manual 3-24*, pp. 5-15, 3-25, 6-21 and 8-16.

30 Rashid, *Taliban*, pp. 102–3.

31 See, for example, George Monbiot's ebullient attack on how US counter-insurgency training was implicated in the work of death squads in Latin America over many decades,

'Backyard Terrorism', *Guardian*, 30 October 2001, p. 17.

32 For a useful account of this general problem (though it does not address the case of Afghanistan), see Joel S. Migdal, *Strong Societies and Weak States: State–Society Relations and State Capabilities in the Third World* (Princeton, NJ: Princeton University Press, 1988).

33 *US Army Field Manual 3-24*, p. 2-1.

34 Review Symposium, 'The New U.S. Army/Marine Corps Counterinsurgency Field Manual as Political Science and Political Praxis', in American Political Science Association, *Perspectives on Politics*, vol. 6, no. 2, June 2008, pp. 347–8 and 350. The four contributions to this symposium are by Stephen Biddle (347–50), Stathis N. Kalyvas (351–3), Wendy Brown (354–7), and Douglas A. Ollivant (357–60). The symposium is available at http://www.apsanet.org/imgtest/POPJune08CounterInsurgency2.pdf.

35 *Ibid*., p. 348. See also the excellent contribution of Stathis N. Kalyas, who argues (p. 352) that by adopting the people's-war model, the authors of the manual assume that the population interacts either with the government or the insurgents. This leads them to conclude, incorrectly, that if the insurgents are removed from the equation the people will move closer to the government.

36 *Ibid*., pp. 347–8 and 350.

37 *US Army Field Manual 3-24*, pp. 1-9 and 7-6. These brief references to Afghanistan do not describe the elements that make the Afghan conflict unique.

38 US Ambassador William B. Wood, Media Roundtable, Kabul, 30 December 2008, available at http://kabul.usembassy.gov/amb_speech_3012.html.

39 Information from a conference at Allied Rapid Reaction Corps headquarters, Rheindahlen, 27 June 2007.

40 US Army officers have been particularly vocal in expressing their concerns about the performance of the US Air Force regarding such matters as bombing missions gone wrong and insufficient priority to the provision of surveillance aircraft. Thom Shanker, 'At Odds with Air Force, Army Adds its own Aviation Unit', *New York Times*, 22 June 2008.

41 Human Rights Watch, *"Troops in Contact": Airstrikes and Civilian Deaths in Afghanistan* (New York: Human Rights Watch, September 2008), p. 2, available at http://hrw.org/reports/2008/afghanistan0908/index.htm.

42 President Hamid Karzai, interview published in *New York Times*, 26 April 2008.

43 Based on conversations I had with Bernard Fall and material in his book *Street Without Joy: Indochina at War, 1946-54* (Harrisburg, PA: Stackpole, 1961). He alludes to these issues in *Viet-Nam Witness 1953-1966* (London: Pall Mall Press, 1966), p. 9. See also his widow's remarkable preface in *Last Reflections on a War* (Garden City, NJ: Doubleday, 1967), pp. 9–10.

44 Astri Suhrke, 'Reconstruction as Modernisation: The 'Post-Conflict' Project in Afghanistan', *Third World Quarterly*, vol. 28, no. 7, 2007, p. 1301.

45 Information from three UNHCR sources: *2006 UNHCR Statistical*

Yearbook (Geneva: UNHCR, December 2007), p. 36; 2007 *Global Trends: Refugees, Asylum-seekers, Returnees, Internally Displaced and Stateless Persons* (Geneva: UNHCR, June 2008), pp. 8 and 9; and the UNHCR Statistical Online Population Database at www.unhcr.org/statistics/populationdatabase.

46 Adam B. Ellick, 'Afghan Refugees Return Home, but Find Only a Life of Desperation', *New York Times*, 2 December 2008. Figure for internally displaced persons from *Economic and Social Rights Report in Afghanistan – III* (Kabul: Afghanistan Independent Human Rights Commission, December 2008), p. 49, available at http://www.aihrc.org.af/index_eng.htm. See also Internal Displacement Monitoring Centre, http://www.internal-displacement.org.

47 For example Peter Beaumont, 'Afghanistan: Fear, Disillusion and Despair: Notes from a Divided Land as Peace Slips Away', *Observer*, 8 June 2008, pp. 34–5.

48 'The Situation in Afghanistan and its Implications for International Peace and Security: Report of the Secretary-General', UN doc. S/2008/617, 23 September 2008, paras 16 and 18.

49 Ann Scott Tyson, 'Commander in Afghanistan Wants More Troops', *Washington Post*, 2 October 2008, p. A19. McKiernan described Afghanistan as 'a far more complex environment than I ever found in Iraq'.

50 See, for example, the report of the draft National Intelligence Estimate on Afghanistan at http://www.nsnetwork.org/node/1017.

51 Figures for casualties of coalition forces in Afghanistan from http://icasualties.org/oef/.

52 Rory Stewart, 'How to Save Afghanistan', *Time*, 17 July 2008.

53 UN Security Council, 4469th meeting, 6 February 2002, UN doc. S/PV.4469, p. 6.

54 See Paul Gallis and Vincent Morelli, *NATO in Afghanistan: A Test of the Atlantic Alliance* (Washington DC: Congressional Research Service, July 2008), p. 1.

55 Statement adopted at a meeting of the North Atlantic Council, Brussels, 12 September 2001.

56 For a particularly well-informed account of the evolution of the roles of the US and NATO since 2001, see Astri Suhrke, 'A Contradictory Mission? NATO from Stabilization to Combat in Afghanistan', *International Peacekeeping*, vol. 15, no. 2, April 2008, pp. 214–36.

57 Marston, 'Lessons in 21st-Century Counterinsurgency: Afghanistan 2001–07', in Daniel Marston and Carter Malkasian (eds), *Counterinsurgency in Modern Warfare* (Oxford: Osprey, 2008), p. 240.

58 See Adam Roberts, 'Torture and Incompetence in the "War on Terror"', *Survival*, vol. 49, no. 1, Spring 2007, pp. 199–212; and the Amnesty report *Afghanistan – Detainees Transferred to Torture: ISAF Complicity?* (London: Amnesty International, November 2007), pp. 20–30.

59 See, for example, US Secretary of Defense Robert Gates's expression of concern about dual command and control in 'Gates: Afghanistan Command Restructuring Worthy of Consideration', remarks at Texarkana,

Texas, 2 May 2008. Available at http://www.defenselink.mil/news/newsarticle.aspx?id=49769. The coalition of forces acting in support of the Afghan government consists of three basic elements. The first is the Afghan National Army, which has a modest manpower level of about 57,000. The second is ISAF, which now comprises some 51,350 troops from 40 NATO and non-NATO countries. The largest contingents are those of the United States, with 19,950 troops, and the United Kingdom, with 8,745. The third basic element is the force of well over 10,000 troops (almost all of them American) who are part of the US *Operation Enduring Freedom*, which focuses particularly on the counter-terrorist mission in Afghanistan. Information on ISAF troop numbers and areas of operation from http://www.nato.int/.

60 See, for example, Gallis and Morelli, *NATO in Afghanistan*, p. 13.

61 For evidence that Taliban fighters in Afghanistan have learned from the mistakes of the period of Taliban rule up to 2001 see Ghaith Abdul-Ahad, 'When I Started I had Six Fighters. Now I have 500', *Guardian*, 15 December 2008, pp. 1, 4 and 5.

62 Robert Birsel, 'Pakistan Parliament Seen United against Militancy', Reuters report from Islamabad, 23 October 2008, http://lite.alertnet.org/thenews/newsdesk/ISL355611.htm.

63 Such a combined approach, in respect of both Afghanistan and Pakistan, has been advocated authoritatively by John A. Nagl and Nathaniel C. Fick, 'Counterinsurgency Field Manual: Afghanistan Edition', *Foreign Policy*, January–February 2009, http://www.foreignpolicy.com/story/cms.php?story_id=4587.

What is Happening in Pakistan?

Hilary Synnott

Because of the importance attached to military operations to destroy al-Qaeda, Pakistan is often viewed in the context of Afghanistan. As a result, events in Pakistan are usually assessed in terms of their effect on US objectives for *Operation Enduring Freedom* and on the operations of NATO's International Security Assistance Force. Their implications for Pakistan as a whole have been neglected and sometimes poorly analysed. But the well-being and stability of Pakistan is of supreme importance in its own right. This importance needs to be reflected in a sound and coherent strategy towards the country. The formulation of such a strategy requires an informed understanding of Pakistan's history; social, political and economic dynamics; and key personalities and influences. And the strategy needs to be underpinned by resources which are commensurate with the importance of the interests to be protected and advanced. Unfortunately, the customary means of acquiring real-time knowledge of events and personalities – diplomacy, media, commercial and other contacts – have been and still are impeded by the hazardous security situation in much of the country. Without a good understanding, actions taken for tactical purposes may have negative strategic consequences which outweigh the hoped-for benefits, as has occurred on several recent occasions.

The challenges in southwestern Asia need to be considered in three separate but related contexts: Afghanistan, the Afghan–Pakistani tribal belt, and

---

Hilary Synnott is Consulting Senior Fellow at the IISS. As a British diplomat, he was High Commissioner in Pakistan from 2000 until 2003, was the Coalition Provisional Authority Regional Coordinator for Southern Iraq in 2003–04, and had postings in India, Jordan, Germany and France.

Survival | vol. 51 no. 1 | February–March 2009 | pp. 61–80 DOI 10.1080/00396330902749681

Pakistan. In the present conjuncture, Pakistan is arguably the most important of the three. With nuclear weapons and a huge army, a population over five times that of Afghanistan, and simultaneous security, political and economic crises, it now seems less able, without outside help, to muddle through its challenges than at any time since its war with India in 1971.

A troubled past

At the end of the British-dominated colonial era in the Indian subcontinent, India achieved its long-sought objective of independence from the United Kingdom in 1947. But the break with Britain was a secondary objective for Pakistan, whose primary goal was to provide a homeland, separate from the new India, for those Muslims in the region who wished to make use of it. The ensuing arrangement was both bloody and inherently unstable. Vivid memories remain, in both India and Pakistan, of the mass movement of people, family separations and slaughter which caused at least half a million and possibly twice as many deaths. The Kashmir dispute, over which there was conflict immediately after the birth of the two new countries and several more subsequently, is still described by Pakistan as 'the unfinished business of partition', over 60 years after that event.

Although religious considerations were ostensibly the reason for Pakistan's separation from India, they have never proved sufficient either to provide a durable national identity or to hold the country together. The war of 1971 and the secession of East Pakistan – Bangladesh – was the most extreme example of this. But despite successive military defeats and humiliations by India, Pakistan persisted with its attempts to use Islam as a means of nation building. Both the army and elected politicians stoked the Kashmir fire for this purpose and as a means of promoting their own separate interests. Certainly, it served the army well to capitalise on the anomalies around the Kashmir issue, and in this they were assisted by all-too-frequent Indian heavy-handedness, injustice and brutalities.

Despite the loss of its Bengali wing, and leaving aside military and political self-service and its constant differences with its massive neighbour, Pakistan remained fraught with tension as it strove to accommodate its multifarious internal differences. Even the main official national language,

Urdu, is not indigenous but was imported from India in an attempt to forge a common means of communication in a country of many dialects and languages. Punjabis form over 60% of the population and dominate both the army and the economy, which causes resentment among residents of Pakistan's other three provinces. The Pashtuns – a tribal people – have a mother tongue which is linguistically distinct from Pakistan's other main languages, as well as separate traditions and culture (of which more below), with separatism never far from the surface. Sindhis, in the southeast, include semi-feudal agricultural labourers and ostentatiously affluent traders in the port city of Karachi. The desert province of Baluchistan, which has seen several violent insurgencies over the decades including the present one, contains almost half of Pakistan's land mass but is home to not much more than 5% of its population. The country as a whole, made up of disparate ethnic groups and with very low human-development indicators – low literacy, widespread poverty, inadequate public health and education – has had a constant problem with stability.

These inherent challenges have been more than reflected in weaknesses in the country's politics and institutions. There have been military takeovers in almost every decade, and until the most recent elections in February 2008 no elected government had lasted a full term since Zulfikar Ali Bhutto's controversial tenure in the 1970s. The 2008 elections were held under a military president, Pervez Musharraf, who had himself taken power through a military coup and who manipulated polls, the constitution and the body politic for his own purposes.

There have been military takeovers in almost every decade

In their struggle to exercise and maintain power, secular politicians and military leaders alike have made use of and sometimes cooperated with and promoted extreme religious groupings for their own ends, while believing they could control such groups. Like Musharraf before him, current President Asif Ali Zardari has forged political links with religious and other parties with violently militant wings. Before he was deposed by Musharraf, the leader of the opposition Pakistan Muslim League (Nawaz) (PML(N)), Nawaz Sharif, had put forward an amendment to the constitution with a view to making sharia law the supreme law in Pakistan, seemingly with the prime minister – Sharif himself – to be the Emir al-Momineem, Leader of the Faithful, and ultimate arbiter of religious dispute.[1]

Before the present epoch and the global threat of violent militancy, the most significant politicisation of religion for secular purposes in the region occurred under the leadership of the military dictator General Zia ul-Haq and during the Soviet occupation of Afghanistan from 1979 to 1989. Massive Western, especially American, resources were channelled through Pakistan, while Zia ul-Haq promoted a jihad, or holy war, against the Soviet presence in the Afghan motherland. The mujahadeen, or holy warriors, were supported and encouraged as valued allies, supplied and managed by the CIA and Pakistan's Inter-Services Intelligence directorate (ISI).[2]

There were two lasting and damaging consequences of this. First, once the Soviets withdrew from Afghanistan, the mujahadeen remained in being. After ten years of encouragement to fight for what was presented as a religious cause, many of these – Pakistanis, Afghans, Saudi Arabians, Yemenis, Palestinians, Algerians and others – were still fired with zeal. Looking for other opportunities, and at the instigation of the ISI, they transferred their attentions to Indian-administered Kashmir, which witnessed an upsurge of violence in 1989 following rigged elections there in 1988. It was the same mujahadeen, continuing the line from 1979, who were mobilised with the regular army in the notorious Kargil offensive in 1999, and who formed groups that Musharraf steadfastly refused to hound out during the year-long, dangerous stand-off with India in 2001–02. From Pakistan's perspective, the jihadists were valuable paramilitary irregulars who could supplement Pakistani security forces in the event of hostilities. And, rather than try to rein them in, it was deemed better to keep them occupied, first on the eastern border harassing Pakistan's existential enemy, India, and subsequently on the western border, engaged with Afghanistan. More recently, however, it has become clear that such zealots, now supplemented by Chechens and Uighurs, are not susceptible to fine-tuning and are quite capable of turning on their controllers and threatening Pakistan itself and the country's wider interests. The attacks in Mumbai in November 2008, which appear to be linked to the Lashkar-e-Tayiba militant group originally created by the ISI, is an example of terrorism which undermines Pakistan's interests.

Attitudes towards the United States

The second consequence of the 1979–89 period was the impact on Pakistani attitudes to the United States. Pakistan is not alone in its ambivalence towards America: many Pakistanis are quick to criticise the country but eager, if possible, to send their children there to seek a better life. In political terms, successive generations have come to regard the United States as a fair-weather friend, ready to deploy funds and muscle in pursuit of its own interests but neglectful, threatening and vindictive when the balance shifts. Washington was slow to recognise Pakistan after partition, was seen as insufficiently supportive in the 1965 conflict with India, and imposed nuclear-related sanctions in September 1979.[3] After the Soviet Union invaded Afghanistan, making Pakistan part of the front line in the Cold War, the sanctions were waived and Pakistan was deluged with US resources. But US priorities and attitudes again changed after the Soviet withdrawal in 1989. Washington imposed new nuclear-related sanctions in 1990. The screws were tightened in 1998 after India's and Pakistan's nuclear tests and yet further the following year after Musharraf's coup against then-Prime Minister Nawaz Sharif, notwithstanding the popularity in Pakistan of Musharraf's actions at the time. Although there was objective justification for Washington's behaviour in some of these cases, this is irrelevant in terms of public perception in Pakistan and its political significance there.

The tragedy of 11 September 2001 changed the relationship once again. Despite some doubts among his corps commanders, Musharraf made the only choice possible when he finally broke his ties with the Taliban and pledged to become an ally in the 'war on terror'.[4] Because the Taliban's excesses were all too apparent by this time – the group had become deeply unpopular in Pakistan – the UN-backed bombing campaign which started in October 2001 did not cause as great a reaction among ordinary Pakistanis as might have been expected. The publicity surrounding the civilian collateral damage caused dismay in Pakistan, as it did elsewhere, and Western embassies and citizens became targets of terrorist attacks. But this violence did not reflect the general mood of the country and protests incited by the religious parties quickly subsided. Paradoxically, there was a much greater shift of mood against the West, and against the United States in particular, in the run-up to the operations in Iraq in 2003, a country with which Pakistan had no significant ties and where it had few interests. The reasons for this were little different from the objections voiced in other Muslim countries, and indeed elsewhere. But the intensity and near-unanimity of sentiment reflected the fact that, by being an active participant in the 'war on terror', Pakistan was associated with US actions.

Since 2003, and as a result of the difficulties facing the coalition in Afghanistan following the resurgence of the Taliban, doubts have emerged about the coalition's staying power. This has led to fears of a re-run of 1989: if the United States were to pull out, Pakistan might again be left to face the consequences of the violence and instability not only in its neighbour Afghanistan but within the tribal belt and Pakistan more generally. Pakistani authorities maintain that, as a result of their military operations in the Pakistani tribal areas, their security forces have suffered more deaths since 2001 – some 1,200 – than the total of all the coalition countries in Afghanistan. This claim is impossible to verify but not implausible. Pakistanis deeply resent being criticised for not doing enough and especially resent the violations of their sovereign territory by the use of drones and, on 3 September 2008, by US forces on the ground.

Doubts have emerged about the coalition's staying power

There is little doubt that the Pakistani army and the ISI, while under the leadership of General Musharraf, were hedging their bets in Afghanistan and indirectly assisting the Afghan Taliban, especially when members of the group successfully sought sanctuary from US pursuit in northern Baluchistan and the Federally Administered Tribal Areas (FATA). For its part, the US military has been incensed that the Pakistanis are unwilling or unable to prevent their territory from being used as a sanctuary from which to launch attacks upon US forces and have felt justified in taking aggressive action in self-defence. This situation, entirely unsatisfactory on both sides, has driven a further dangerous wedge between two governments which seek to present themselves as allies.

Sadly, the US relationship with Pakistan over the years has become associated with support for military rulers: General Ayub Khan when the US was forging alliances in Asia; General Zia ul-Haq during the Soviet occupation of Afghanistan; and General Musharraf during the 'war on terror'. The promotion of democracy has seemed to have taken second, if not a lower, place. Thus some of the blame for the existing chronic weaknesses of Pakistan's non-military institutions is laid justifiably at the door of the United States.

Domestic politics

Narratives and perceptions surrounding Pakistan's troubled history and its turbulent relationship with the United States, whether or not soundly based, play directly into Pakistani domestic politics. Any strategy towards Pakistan must therefore take full account of these perceptions and, ideally, seek to redress the balance.

Subject to more pressing exigencies, Western governments prefer to support the concept of democracy. After Zia ul-Haq's mysterious death in 1988, Washington had high hopes for the Western-educated Benazir Bhutto, who became prime minister in the subsequent elections. The United States also proved ready to transfer support to her political rival Nawaz Sharif when then-President Ghulam Ishaq Khan dismissed Bhutto some two years later amid allegations of corruption. Although Sharif's party had been largely created by the army as a counterweight to Bhutto, whom the army distrusted, Sharif met a similar fate in a similar timescale. Each contender then served similarly foreshortened terms once again, each of which seemed worse than the former. It was true that elected governments' powers were greatly constrained by the role of the army in the wings, but this could not excuse the seeming inability of both key political parties to live up to their responsibilities as regards both governance and integrity. Elected leaderships signally failed to consolidate democracy while the abuses of their power increased in scale.

Against this background, successive chiefs of army staff were frequently urged by the Pakistani establishment to intervene once again to improve the effectiveness of governance and to reduce self-serving corruption. Three years into Sharif's second term as prime minister, after the Kargil operation and following wholesale mismanagement, the country was nearly bankrupt. In a bid to accumulate more power, Sharif had by this time intervened in the operation of the judiciary and most other national institutions, and had pledged to introduce sharia law. But the final straw proved to be his attempts to meddle in what the army regarded as its own business. In October 1999 then-Army Chief Musharraf presided over a bloodless coup. This was strongly condemned by the US and British governments, but Sharif's departure was widely welcomed in Pakistan. Western opposition moderated as Musharraf started to introduce some overdue and seemingly beneficial reforms, made use of his presentational skills to project his well-meaning intentions, and presided over impressive but unsustainable economic growth.

But with the passage of time Musharraf became less able to maintain the momentum of reform; he had some damaging setbacks, made major errors of judgement, engaged in transparently dubious political expediencies, pursued unsound economic policies and increasingly lost popular support. His participation in the United States' 'war on terror' was denounced as 'fighting Washington's war', and he was blamed for allowing the nation's honour to be humiliated by US criticism that he was not doing enough. Nonetheless, the US government appeared increasingly to be pinning their hopes on Musharraf personally, evidently fearing for the Pakistani role if he were to go.

In a desperate attempt to maintain his position, Musharraf sacked Pakistan's chief justice in March 2007 and suspended some 60 judges. The ensuing militant protests by the judiciary, being neither religious nor party political activists, introduced a new and unusual dimension to the political crisis. When Musharraf rashly declared a state of emergency on 3 November, it became clear that his days were numbered.

The US and British governments, concerned to protect their interests in Afghanistan and in countering terrorism, still hoped they could promote democracy while maintaining Musharraf in power. Despite the deep mutual antipathy of the two protagonists, they had been actively encouraging Benazir Bhutto and Musharraf to come to an accommodation. Musharraf was induced to arrange for the criminal charges against Bhutto and her husband to be dropped so as to allow her to lead her Pakistan People's Party (PPP) in the general elections due in late 2007. The plan was that, with the aid of the so-called 'king's party', the Pakistan Muslim League (Quaid-i-Azam) (PML(Q)), which backed Musharraf, the PPP would assume the prime ministership while Musharraf remained as president, having first stood down as chief of army staff. Sharif, however, would not be allowed to return to Pakistan from his forced exile in Saudi Arabia, which would severely disadvantage his Pakistan Muslim League (Nawaz) (PML(N)), Pakistan's only other significant national political party, which might otherwise threaten Bhutto's ascendancy.

But the turmoil surrounding the judiciary renewed Sharif's confidence and, after one failed attempt in September, he too returned to Pakistan in

November, although this gave him little time to mobilise his support and allocate seats for his party's electoral candidates. Bhutto's assassination on 27 December 2007, soon after the lifting of the state of emergency, was a major political upset, throwing awry the careful plans for the elections as well as the US–UK machinations. The new army chief, General Afshaq Kayani, ordered the army to stand aside from politics.

As a result, the postponed elections in February 2008 included both main national parties and were relatively fair and open. Bhutto had bequeathed to her husband, Asif Ali Zardari, the leadership of the PPP, which gained a plurality, but not an overall majority. Sharif's PML(N) came in second. Neither Sharif nor Zardari stood for election themselves. The PML(Q) and the religious parties, most of which boycotted the elections, lost heavily.

Sharif's decision to enter into a coalition with the PPP was widely welcomed as an apparent recognition of the need to work together in a time of crisis. But the coalition fell apart acrimoniously in August 2008 when, under threat of impeachment, Musharraf stood down as president and Zardari, possibly fearful that he might be prosecuted for past corruption, refused to reinstate the chief justice whom Musharraf had sacked.

The coalition fell apart acrimoniously

Zardari nonetheless managed to secure the necessary majorities in the national and provincial assemblies to be elected to the now-vacant presidency. Thanks to the measures taken by Musharraf to strengthen the office of president, Zardari's position under the current, heavily amended constitution is therefore very strong, not least because there is no way to remove a president before the end of his term so long as he has sufficient support in parliament. But in other important respects Zardari's position is weak. Opinion polls reveal him to be deeply unpopular with the general public as a result of sustained and long-standing allegations of corruption and even murder. He has caused tension within his own party by sacking many of his wife's former advisers and placing his own associates in key positions. He is unpopular, too, within the army because of his failure to consult before announcing major changes of policy and practice, several of which the army caused to be immediately reversed. And his perceived association with the United States, to which, through its support for his wife, he owes his position, risks being used to his political disadvantage.

Two recent developments have had a particular political impact in Pakistan. A military operation involving US ground forces in the South Waziristan Tribal Agency on 3 September 2008, the second day of the holy month of Ramadan, caused an immediate outcry, with strong condemnation by both houses of parliament, the prime minister, the chief of the army and finally by Zardari himself.

Secondly, when Zardari was in Washington at the end of October 2008, he unprecedentedly described the militants in Kashmir as 'terrorists', rather than using the hallowed terminology of 'freedom fighters'. Welcome as this may have been in the US administration, it is regarded within Pakistan as a capitulation, made all the worse by the fact that Zardari returned from Washington empty-handed. The impression therefore remained that the United States was unwilling to help Pakistan with its economic and social problems, greatly exacerbated by rising food and fuel prices and the fall of the currency, and was solely preoccupied with military and security issues in and around the tribal belt.

The Pashtuns

It is simplistic to describe the militants in Afghanistan and Pakistan as 'al-Qaeda and the Taliban', as if the coalition's opponents were members of two essentially homogenous groups which, whatever their differences, share similar aims. In practice the composition and motivations of the fighters and of their sympathisers and supporters are varied, complex and change over time.

After 11 September 2001, Pakistani attitudes towards al-Qaeda and other non-indigenous fighters have been fairly straightforward, even though intermarriage with Uzbeks and Tajiks since 1979 has blurred distinctions. Pakistani security forces have proved ready to attack and kill 'foreign fighters' without hesitation and to significant effect.

The approach towards the Taliban and others has, at least until mid 2008, been carefully nuanced and calibrated. The Taliban, or religious students, who recognise the religious and secular leadership of Mullah Omar, went to ground after sustained US attacks in the aftermath of 11 September. But they have re-emerged in a new and more effective form, partly because the coalition's initial concentration on Kabul and the United States' preoccupation with Iraq eased much of the pressure on them. Many Taliban are indeed motivated by a radical religious zeal arising from a form of Islam which many Muslims regard as a distortion of the faith. But there are many others in the fight: criminals, warlords, narcotics smugglers, young men who see little other future for themselves, and insurgents who object to the presence of foreigners engaging in violence in their homeland.

Almost all of the Afghan and Pakistani Taliban are Pashtuns. Some understanding of the Pashtuns (or Pushtuns, Pukhtuns or Pathans) is therefore essential in addressing the threat to external interests. Hardened by millennia of conflict in the exceptionally rough terrain of what has been the transit route for waves of invaders from Alexander the Great to the British and Soviets, the Pashtuns are tough, warlike and fiercely proud of their religion, heritage and traditions. Figures are inaccurate but, of the roughly 40 million people in the region, perhaps 15m live in Afghanistan, while the rest are mainly in Pakistan's FATA (3.5m[5]), the North West Frontier Province (NWFP), northern Baluchistan and the city of Karachi. The boundary between Pakistan and Afghanistan, the 1,500-mile-long Durand Line drawn up by the British in the late nineteenth century, divides tribes and villages and has never been accepted by Afghanistan as an international border. Thousands of individuals pass across the line every day, which has been blurred by the presence of millions of Afghan refugees in Pakistan and, in recent months, by the movement of 20,000 refugees from Bajaur Agency in Pakistan to Afghanistan. The terrain makes it impossible in practice to seal off the border and the long-established trade and smuggling routes which cross it, and it is therefore relatively easy for fighters or fleers to move in either direction. Pakistani security forces cannot control them, even if they were inclined to do so, despite some 1,000 border posts on the Pakistani side and a smaller number to the west.

Almost all of the Taliban are Pashtuns

Of greater political significance than the physical ease of passage across the border is the tribal code, or Pashtunwali. Although it has been severely distorted in recent decades by the influence of criminality, the effects of the narcotics trade and the intrusion of the Pakistani army, it continues to exercise a powerful influence. Traditionally, conformity to this code is expected of any Muslim living in the tribal area. Pashtunwali centres on the concept of honour (*izzat*) of the individual or the tribe, from which spring the obligations of revenge and sanctuary (*badal* and *melmastia*). There are also mechanisms for mediation, and rules to govern ostracism or, in extreme cases, death, if the code's obligations are not met. Historically, in their attempts to manage these areas (which they found impossible to settle), the British used a mixture of inducements and retribution, employing the concepts of collective responsibility and collective punishment in line with the pre-existing norms of Pashtunwali. In practice, when the British used the destruction of property as a form of such punishment, they gave prior notice so as to avoid loss of life.[6]

When, in the 1980s, Washington encouraged the mujahadeen to drive out the Red Army from Afghanistan, they were employing the same concepts of Pashtunwali to deal with an uninvited infidel who had violated Pashtun honour by entering the tribal homelands without invitation. It is evident that this same code applies similarly to coalition operations in Afghanistan and the FATA, and that it is also relevant to tribal behaviour towards the coalition's opponents. The coalition's presence, like that of the Soviet Union and the British, is seen as a violation of communal honour which the code and custom require to be avenged. The deaths of tribesmen at the hands of coalition forces must similarly be avenged, if not by the closest male relative then by his descendants. A tribesman fleeing from the coalition who seeks refuge must be given hospitality and safe passage: the greater the threat, the greater the honour accruing to the host.

The situation is rendered more complex by the fact that the seven tribal agencies and the six frontier regions which make up the FATA have a semi-autonomous status within the Pakistani constitution that reflects their ancient customs. This special status was pledged to them by Mohammed Ali Jinnah when they were induced to join the new Pakistan in 1948, and severely curtails the writ of federal authorities and security forces.[7] It remains fiercely guarded by the tribal people in the FATA.

For such reasons, until about 2004 the Pakistani authorities used the paramilitary Frontier Corps, almost all Pashtuns themselves but led by Pakistani army officers, in any operation in the tribal belt, rather than the Punjabi-majority regular army. The large-scale intervention of the army in 2004, quite unused to conducting counter-insurgency operations, was bitterly resented within the FATA, and has been violently opposed.

In August 2008, faced with increasing disorder, Pakistani armed forces radically stepped up their operations in the Bajaur Tribal Agency, using the regular army on the ground, heavy artillery, US-supplied weaponry, and heavy bombing by Air Force F16s, also supplied by the United States. (Because Pakistan's military is unable to engage on several fronts at once, uprisings in the 'settled area' of Swat continue largely unchecked.) Following the customary practice, the armed forces called on local inhabitants to give up foreign fighters and to leave the area themselves. At the time of writing, these and other operations in NWFP have killed over 1,000 people, and estimates of the total number of displaced persons vary from 200,000 to 700,000.[8] Several thousand have fled to Afghan territory. Although a significant number of foreign fighters have been killed, the use of non-precision weaponry and the army's lack of experience in counter-insurgency techniques may be expected to have considerable long-term political consequences. Nor is it clear who it is that the army is fighting or whether Pakistan hopes to target foreigners and the Pakistani Taliban while allowing the Afghan Taliban and 'freedom fighters' greater impunity. The army's opponents, however, may come from a wider spectrum, which includes those with any grievance against them.

There are other unwelcome phenomena which have arisen in the rest of Pakistan in recent years. Before 2005, four years after the start of the coalition's operations in Afghanistan, suicide bombings in Pakistan were very unusual, with scarcely a handful each year. But the number rose to 57 in 2007 and 63 in 2008.[9] A 'Pakistan Taliban', the Tehrik-e-Taliban, has emerged in South Waziristan Tribal Agency, and Pakistani authorities say the leader of this group, Baitullah Mehsud, is one of the main instigators of such bombings. Terrorists are now demonstrably able to operate throughout the country, including in the heart of the capital, Islamabad, as was shown by the defiance of religious militants during the siege of the Red Mosque in July 2007, during which over 100 were killed by Pakistani security forces, and the bombing of the Marriot Hotel, close to the parliament buildings, in September, when suicide bombers killed over 50.

There are, however, some more positive signs. In recent months Pakistani authorities have taken long-overdue steps against the presence of the *shura*, or advisory council, of the Afghan Taliban in the western Pakistani city of Quetta. There is good reason to believe that there are serious differences between Mehsud and Mullah Omar, leader of the Afghan Taliban. And it may be that al-Qaeda's presence is no longer welcome to the Afghan Taliban, although rumours to this effect are difficult to verify. But it would be wishful thinking to expect any significant reduction in the threat to coalition interests. Of greater concern is the increase in numbers, diversity, sophistication and ferocity of the coalition's opponents.

Assessment and future actions

Despite these considerable difficulties it would be wrong to regard Pakistan as a lost cause or as doomed to go down the path of violent extremism, leading to the establishment of an Islamic caliphate. In this context, Pakistan's diversity is also one of its strengths. Although the vast majority of Pakistanis are Sunni Muslims, there are minority Shia and Ismaili communities as well as Christians. There is also diversity among the Sunni, which encompasses the Deobandi and Barelvis sects. Far from being sympathetic to the cause of radical Salafists and Wahhabis, many Pakistani Muslims have been influenced by the Sufi tradition of Islam, which is unaggressive and tolerant, and enriched by poetry, song and dance. The great majority of the much-maligned madrassas, or religious schools, fulfil an essential social service by providing food, clothing and shelter to poor children, as the state's primary-education system has been severely weakened through neglect and corruption by successive governments. It is a recalcitrant minority of such schools which actively support the Taliban and which need firmer control.

Nor is Pakistani nuclear weaponry likely to fall into terrorist hands. The Pakistani military remain a powerful and cohesive force. And, with some external help, it has established powerful security measures.

It is therefore reasonable to judge that, while violence and disorder will very likely increase unless or until greater attention is paid to Pakistan's challenges, it is unlikely that the country is on the brink of state failure. Some parts of the country undoubtedly suffer from severe security challenges and deprivation. But others, including much of Punjab, provide a core of stability which has weathered many storms in the past and which is unlikely to disintegrate into chaos.

It is unlikely that the country is on the brink of state failure

In order to offset the negative effects of the coalition's military operations in Afghanistan on Pakistan's own security, Pakistan's stability and its perceptions of the West might be improved by doing more to help Pakistanis cope with the other significant challenges that affect their daily lives. Since 2001, the bulk of Western financial and other assistance has been focused on security-related sectors.[10] The social sector, sorely neglected, is in desperate need of capacity-building. An increase of resources for this purpose, to a level similar to that devoted to the security sector, would, if carefully directed and administered, provide new opportunities for the Pakistani people and do much to counter the pernicious belief that they are suffering as a result of fighting someone else's war.

There can, however, be no expectation of quick results or a rapid transformation of the security situation in the tribal areas: the wounds are too deep and Pakistan's institutions too weak to be able to deliver progress swiftly. In the beginning, therefore, a reasonable aspiration would be to avoid making matters worse, even if this may seem unambitious and require the tacit acknowledgement that mistakes have indeed been made. Ill-judged US military operations on the ground, like that on 3 September, appear to have been suspended. But there should be no doubt that, were they to be resumed in similar circumstances, Pakistani security forces could well retaliate against the 'invaders', which could have devastating political consequences in the country.

Both the avoidance of political error and the achievement of any progress will require a strategy towards Pakistan that takes account of the country's management and governance as a whole. The first requirement, therefore, will be for the United States to develop and then publicise an approach towards Pakistan which is essentially benign and designed to help across the broad range of challenges described above. But this should not take the form of blank cheques for unmonitored and uncontrolled expenditure. Nor would the security situation allow for any significant deployment on the ground of Western development administrators who, in traditional development-cooperation relationships, would normally monitor projects and programmes, control expenditure and ensure good practice. This points to a need to develop imaginative and enhanced roles for the international financial institutions, who might be better placed to recruit and deploy indigenous administrators, and to craft contracting and accounting mechanisms which may be more manageable in the local environment than the strict rules normally required by donor governments.

Legislation sponsored by US Senators Joe Biden (before he became vice president) and Richard Lugar, currently before the US Congress, envisages the provision of $1.5 billion per year for five years, with a possible extension, to be devoted to non-military sectors.[11] If enacted, this would undoubtedly help. But it is unlikely to be sufficient.

In short, it is necessary to demonstrate that every effort will be made to bring about improvements to ordinary Pakistanis' well-being, and that these efforts will be sustained and not subject to short-term political expediencies or personalities. Counterpart efforts will be equally essential, if any such partnership is to be truly effective, to improve the transparency in the use of donors' funding. Priorities should include strengthening areas of governance such as the judiciary and the electoral system, and enhancing social and economic sectors including employment-creation, education, primary healthcare, water management and storage, and energy supply and management.

At the same time and in the light of past experience, friends of Pakistan would do well not to be too prescriptive in dealing with Pakistan's complex political scene. If Zardari's political star were to fall, the current working assumption is that his leadership, in some form or other, would be replaced by that of Nawaz Sharif. In view of the failings of Sharif's previous leadership, there are grounds for concerns about such a prospect. Responding to popular sentiment, he may be expected to be cautious about alignment with the United States and, as in the past, have closer relationships with religious groupings than does the centre-left PPP. Nor can there be any certainty that he has learned from the mistakes of the past. But that is equally true of Zardari, about whom there are other grounds for concern. For all his past and possibly present shortcomings, Sharif is no revolutionary and may be expected to adopt a pragmatic approach. His stated priorities for the issues which need attention and reform exhibit a clearer focus than in the past and offer some hope that his leadership would extend beyond mere politicking and the maximisation of power.

It is also tempting to hope that the two leaders might come to recognise that the seriousness of the challenges Pakistan now faces call for them to work round their differences. Even if they proved unable to re-form a coalition, it could be hugely beneficial if they were to cooperate over issues on which they have a broad measure of agreement and make use of the totality of the talent and experience which exists within their respective parties. It would indeed be a mark of statesmanship if each were to give the lie to the general assumption that Islamabad is not big enough for both of them.

Notes

1 Owen Bennett-Jones, *Pakistan* (London: Yale University Press, 2002), p. 18; M. Ziauddin, 'Bill Has to Cover Two More Stages', Pakistani.org, 10 October 1998, http://www.pakistani.org/pakistan/constitution/amendments/15thamendment_stages.html.

2 Ian Talbot, *Pakistan: A Modern History* (New York: Palgrave Macmillan, 2005), pp. 267–70.

3 *Ibid.*, p. 249.

4 Pervez Musharraf, *In The Line of Fire* (New York: Free Press, 2006), pp. 201–7.

5 Details set out in a paper presented by Lt-Gen. Tariq Majid at the IISS Shangri-La Dialogue, 31 May 2008. The 1998 census records a total population of 3.176 million. The higher figure of 3.5m is commonly used for the current population. See 'Population Estimates by Agencies/ Frontier Regions of Fata 1981 to 2002', Bureau of Statistics, Planning & Development Department, Government of NWFP, http://www.nwfpbos.sdnpk.org/fds/2000/10.htm

6 Louis Dupree, *Afghanistan* (New Delhi: Rama Publishers, 1980), p. 126.

7 Article 1 declares that the FATA are part of Pakistan. Article 246 defines the Tribal Areas, of which the FATA are a part. Article 247 sets out how the

Tribal Areas should be administered, placing considerable limitations on the role of the federal parliament and federal judiciary, although it also provides for the possibility of some changes of practice.

8 'Afghanistan–Pakistan: Bajaur Residents Flee Fighting', UNHCR, 25 September 2008, http://www.unhcr.org/refworld/country,,,,PAK,,48f2f6161a,0.html.

9 Ahmed Rashid, 'Jihadi Suicide Bombers: The New Wave', *New York Review of Books*, 10 June 2008, http://www.nybooks.com/articles/21473; Hasan Askari Rizvi, 'A Difficult New Year', *Daily Times*, 4 January 2008.

10 'Major U.S. Arms Sales and Grants to Pakistan Since 2001', prepared for the Congressional Research Service by K. Alan Kronstadt, Specialist in South Asian Affairs,28 August 2008.

11 'Biden, Lugar introduce New Legislation on Pakistan', Press Release, 15 June 2008, http://biden.senate.gov/press/press_releases/release/?id=7C1A5AA4-D35B-4A8E-8D8E-3B3B156099D5.

Afghan Q&A: Robert Gates

François Heisbourg, Chairman, IISS Council:

My question is about Afghanistan. You described the challenges and some of the responses, but I would like you to go into more detail about what you expect other nations to contribute in military terms. Since I am a European, what would you expect European partners and allies of the United States to be contributing? More generally, what would you consider to be the appropriate balance, the appropriate policy mix, between the military surge on the one hand, and the political and economic efforts in Afghanistan and in the region on the other hand?

Robert Gates, US Secretary of Defense:

One of my European colleagues remarked last year that I was guilty of megaphone diplomacy when it came to seeking additional help. At this point, US forces and the forces of our partner nations are about in balance. There are approximately 33,000 US troops, and approximately 33,000+ troops from Europe and other partner nations. As is evident to everyone who reads the newspaper, we are looking at sending significant additional forces over the course of the next months; another brigade combat team will go in January, and we hope to send two more by late spring or early summer.

With respect to the Europeans and our other partner nations, it is important to recognise the achievement. The Europeans and our partners have actually increased their forces by about 10,000 over the past year, and that is an important achievement. I think it is also important in terms of fairness to remember that many of these nations also have deployments under the UN and other charters in places like Darfur, the Balkans and elsewhere. People are doing their best. But, the reality is that the European members of NATO have approximately 2.5 million people under arms, not including the United States, and so I must admit to some frustration in trying to get some few thousand more to help us in particular to train the Afghan National Army and National Police. Ultimately, it is the Afghan forces who are everyone's ticket out of Afghanistan, and for the long-term success of Afghanistan.

This is the Afghans' fight. It is their country, their fight and their future that is at stake. I believe one of the highest priorities is to get the resources into the country to help train, accelerate the expansion of the Afghan security

5th IISS Regional Security Summit, The Manama Dialogue, First Plenary Session, 'The US and the Regional Balance of Power'. For transcripts of the speech and Q&A session see http://www.iiss.org/conferences/the-iiss-regional-security-summit/manama-dialogue-2008/plenary-sessions-and-speeches/.

Survival | vol. 51 no. 1 | February–March 2009 | pp. 81–82 DOI 10.1080/00396330902749699

forces, and to get more of them into the fight in their own country. We need to make sure that we find ways to keep the Afghans up front. We are there to help them. We are not there as occupiers. We are not there necessarily for our own interests. We are also there to help the Afghan government.

I do worry. I am prepared to go forward with the forces that the commander has requested during 2009, but I do worry, ultimately, about the size of the footprint of foreign forces on Afghan soil. Having been engaged in the fight to expel the Soviets from Afghanistan more than 20 years ago, I am more mindful than most that, with 120,000 troops, the Soviets still lost. This is because they never had the support of the Afghan people. We need to give serious consideration as to how many foreign forces we should send in. After we complete these troop increases that we are talking about, we ought to think long and hard about how many more go in, whether they are from the US or Europe or anywhere else.

The reality is that we have to help the Afghans get onto their feet. This is where the economic side comes in, and the development side. Afghanistan is a desperately poor country. It is the fourth or fifth poorest country in the world. We are all going to have to be in Afghanistan for a very long time. A good example is the contrast between the government revenues of Iraq, which this year will be approximately $60–70 billion, and the annual revenues of the Afghan government, which will be approximately $700m. The sustaining cost of an Afghan National Army of 134,000 people is approximately $2bn a year. This is roughly three times the government's revenues. We are all going to have to be there for a long time.

Many countries are engaged in trying to help with economic development and the development of institutions in Afghanistan. My biggest concern is the lack of coordination of those efforts in Afghanistan. I am a huge admirer of Ambassador Kai Eide. I believe that, regrettably, Kai has not been given the tools that he needs to do the job by the UN: the money and the people. We also need to do more to help him in terms of providing him with information with what each of us is doing; in terms of development projects and the money we are spending on roads, and so on. He will then be able to evaluate what is working and what is not working, and help coordinate efforts.

The European emphasis on a comprehensive strategy that combines both security and economic development, and civic development, is exactly right. We have to recognise that we cannot win this fight with military power alone. The civil side has to be successful. This is a very long-winded answer to your question, but the bottom line is that we will be sending in some more forces. However, the real focus needs to be in keeping the Afghans up front. This is for the Afghan people. We must be seen as their ally and their friend, and not as occupiers. Finally, we have to sustain the economic development and institution-building in Afghanistan at the national, provincial and district levels.

Forum

The Way Forward in Afghanistan: Three Views

End the War on Terror

Barnett R. Rubin

The situation in Afghanistan has turned so far against the United States, NATO, the international community, and those Afghans who originally hoped that the post-11 September 2001 intervention would finally bring them a chance for normal lives, that it will be very difficult to salvage. Al-Qaeda has established a new safe haven in the Federally Administered Tribal Areas of Pakistan, from which it supports insurgencies in Afghanistan and Pakistan and continues its global planning against the United States and its allies. Its press releases are so frequent that they are hardly newsworthy unless they feature video of Osama bin Laden himself. Negative trends in Afghanistan include the deterioration of security, Afghan governance and regional stability. The stability of Pakistan, a nuclear-weapons state that has been the main source of proliferation over the past two decades, is now at serious risk. Rising India–Pakistan tensions further exacerbate the regional risk, as do tensions over Iran's nuclear programme and its relations with Hizbullah and Hamas.

The task in Afghanistan would have been difficult under any circumstances. The Bush administration's unique record of incompetence, fecklessness and criminality has assured that the Obama administration

Barnett R. Rubin is Director of Studies and a Senior Fellow at New York University's Center on International Cooperation and the author of *The Fragmentation of Afghanistan* (2nd ed., 2002) He served as an adviser to the Special Representative of the Secretary-General at the UN Talks on Afghanistan in Bonn in 2001.

 DOI 10 1080/00396330902749707

inherits its responsibilities under the worst possible circumstances, not only in the region, but globally as well. Still, as President Obama's chief of staff Rahm Emanuel said of the economic situation, 'You never want a serious crisis to go to waste'.

This serious crisis may finally force equally serious thinking about the goals of the international intervention in Afghanistan and the means required to have any serious hope of attaining or approaching them. Rather than proclaim objectives limited only by the audacity of our imaginations (an Islamic democratic, stable, gender-sensitive and prosperous Afghanistan) and the paucity of our means (fewer resources per capita than any other such operation), we need to align objectives with reality, and means with objectives.

The most important change in the definition of US objectives is to explicitly renounce the 'war on terror'. Instead the United States is engaged in a war against al-Qaeda, which attacked America and its allies. Al-Qaeda, a non-territorial transnational network, can obtain a safe haven only through alliance with groups such as the Taliban, which have a national or ethnic base connected to a territory and population. Such alliances are inherently unstable, however, in so far as any territorialised political movement has objectives related to the territory and population where it is based, objectives which are necessarily different from al-Qaeda's global goals of re-establishing the Islamic caliphate throughout Muslim territory.

The 'war on terror', which amalgamated all Islamist groups that used violence into a common threat, strengthened its primary target, al-Qaeda, by creating incentives for local groups treated as 'terrorists' to ally themselves with al-Qaeda. All handbooks of war, dating back at least to Sun Tzu, have recommended dividing the enemy. The 'war on terror' did the opposite.

While counter-terrorism requires military and intelligence tools, only a drastic strategic reorientation can provide those with their required political complement. In the Afghan context, such a clear, public reorientation of counter-terrorism policies should lead the United States and its partners in Afghanistan to offer political negotiations to any Taliban and other insurgents who are willing to separate themselves from al-Qaeda. Such a policy has been in effect formally for several years, but related policies on sanctions, detention and reintegration have not been restructured to reflect that stance. Political accommodation with groups that accept effective guarantees against the creation or protection of terrorist sanctuaries will require reciprocal US guarantees against detention or sanctions for any leader willing to enter into such an agreement. Thus far the United States has no mechanism to assure that such a guarantee is observed by the multitude of agencies involved in the counter-terrorism effort.

The same shift in counter-terrorism policy should apply to Pakistan, though it will take a different form. The United States should support efforts by the elected government of Pakistan to separate Pakistani insurgents from al-Qaeda and other foreign fighters, in particular by supporting programmes to reform the status of the Federally Administered Tribal Areas to address the grievances and isolation of the population there.

Separating Afghan or Pakistani Islamic insurgents from al-Qaeda would constitute a serious political setback for the latter that would damage its claims to legitimacy and its recruitment capacity in the Islamic world. Much of the diffuse international sympathy for al-Qaeda (now on the decline) derives from resistance to 'occupations' of Afghanistan and Iraq. Any political settlement with Afghan insurgents, especially the Taliban leadership, would deprive al-Qaeda of that claim.

Inclusion of the Taliban leadership and other insurgents in a political settlement does not mean returning Afghanistan to Taliban rule or abandoning the broad portion of the Afghan political spectrum that has worked with the international community and welcomed liberation from Taliban rule. Nor is it meant as a quick fix to replace policies aimed at the regional factors behind the insurgency or the corruption and abuse that have so weakened the Afghan government. A political settlement cannot succeed without policy changes by the Taliban's regional sponsors, and insurgents cannot be reintegrated unless the government becomes more credible.

What the United States should ask of its Afghan partners is that any political agreement be based on recognising the authority of the Afghan government and its security forces throughout the territory of Afghanistan. Participation in power among (more or less) disarmed political groups through coalition or cooptation is acceptable; division of the country into spheres of influence under the control of multiple authorities or security forces is not. Power sharing in the latter sense permits formation of safe havens.

This is what the US and Afghan governments should mean when they state that negotiating partners must accept the Afghan constitution. This should not mean passage of an ideological test requiring agreement with every article but recognition of the sovereignty of the government established by the constitution. Many issues dealt with (often ambiguously) by the constitution will remain contentious for a long time, and not only to insurgents. Insurgents who lay down arms will have the same rights as other Afghans to disagree with and seek to change the constitution through peaceful means.

Closing Guantanamo is a first step

Such a declaratory policy is already in effect, but no one takes it seriously, since the existing policies on detention and sanctions send the other message. Taliban and al-Qaeda are detained together in Guantanamo and sanctioned together by the UN Security Council. Closing Guantanamo is a first step. Afghan and Pakistani detainees (except for those closely linked to the al-Qaeda leadership, a category which includes no Afghans) should be transferred to national custody or released. The international community will have to fund generous reintegration packages in both countries.

Both national and international sanctions regimes should be changed to guarantee security and integration of insurgents who join the political process. Russia has thus far opposed removal of anyone from the sanctions list for reasons that should be explored further. The main reason is probably its concern that the purpose of integrating former Taliban is to consolidate a NATO base in its near abroad. Diplomatic efforts to overcome these objectives could serve common Russian and Western interests in the elimination of the threat from al-Qaeda.

Such a policy change will not work by itself. To succeed it must be accompanied by military, security and governance efforts that enable the Afghan government to present a more credible alternative than it has. The Afghan Taliban leaders are dependent on their Pakistani sponsors and supporters, including in the country's military and security apparatus, for their safe haven, and regional diplomacy aimed at changing Pakistan's security calculus remains essential. The core of such a policy is firm support for the efforts of the elected government of Pakistan to gain control of the country's security policy and define the national interest as the welfare of the citizens of Pakistan.

Within Pakistan, integrating the Federally Administered Tribal Areas into what Pakistanis call the 'mainstream' is also not a quick fix. It will require a strategy that will take many years. There will be armed resistance by al-Qaeda and many other groups whose existence depends on the isolated nature of these areas. But gaining control of national territory in order to protect the rights of Pakistani citizens will certainly provide a more legitimate mission for the country's security forces than assisting the United States in its 'war on terror'.

No single policy change can solve any problem, let alone a set of problems so complex and interdependent as those of this region. But such a bold, clear announcement, followed by concrete public steps can go a long way toward transforming the poisonous environment we all have inherited.

What Future for Afghanistan?

Amin Saikal

Afghanistan is in the grip of long-term, violent, structural disorder and insecurity. Senior NATO political and military figures have voiced strong scepticism about winning against the Taliban and their supporters, and have intimated that the United States and its allies should focus more on generating the necessary conditions for security than on democracy. Some have supported Afghan President Hamid Karzai in his efforts to negotiate with the Taliban to produce a viable settlement. The Karzai government and its international backers are not as yet in a position to bargain for such a settlement. They must first not only impress upon the Taliban that the insur-

Amin Saikal is Professor of Political Science and Director of the Centre for Arab and Islamic Studies (the Middle East and Central Asia) at the Australian National University, and author of *Modern Afghanistan: A History of Struggle and Survival* (2006) and *The Rise and Fall of the Shah: Iran from Autocracy to Religious Rule* (2009).

gency cannot succeed militarily, but also address the political and strategic vacuum their own failures created and the Taliban have exploited.

The strong presidential system established under Karzai has not worked and is in urgent need of restructuring. It has proven unsuitable for a traditional, conflict-ridden Muslim country with myriad ethnic, tribal, linguistic and sectarian divisions. Such a system typically produces a single winner and many disgruntled losers capable of challenging and undermining the victor. It places too much responsibility with one person, leaving the president highly vulnerable to unrealistic public expectations and discontent, especially when things go wrong. Yet, paradoxically, the actual powers of the president are often less than they appear on paper. To manage the affairs of the state the president may find it tempting to engage in a politics of patronage and nepotism. This is what has happened in Afghanistan.

Karzai has presided over an increasingly corrupt and dysfunctional government. He has not been able to create a united and competent governing elite; and personal friendship, family, tribal, ethnic and factional connections rather than merit have formed the basis for most senior government appointments. He has surrounded himself with many political and ethnic entrepreneurs, most of whom come from the Afghan diaspora and lack the qualifications and experience to serve Afghanistan beyond their individual interests. Nor has Karzai fulfilled the expectations of a majority of the Afghan people for peace, security and improved standards of living.

The opposition elites, most importantly the Taliban and its allies – including the Hizb-e-Islami of the former maverick mujahadeen leader, Gulbuddin Hematyar, as well as foreign intelligence agencies – have penetrated the government and institutions at many levels. This has not only compromised governmental activities, but also held the United States and its allies back from channelling most of their reconstruction aid through the government and coordinating closely with it on major policy and security operations. The result has been a massive political and security vacuum, enabling the Taliban and its supporters increasingly to appear more credible than the government.

Afghanistan needs a diversified, party-based parliamentary system of governance, headed by a prime minister, who would come from the parliament with a parliamentary majority, under a figurehead president, elected by the parliament and provincial assemblies. Such a system could be more inclusive, providing for a range of influential actors to be locked in positions of national obligation and responsibility. It could also provide more accountability, transparency, efficiency and popular connection to the political system. To achieve this, the Afghan Constitution of 2004 would need to be substantially modified, which in turn requires convening a new Loya Jirga (grand assembly) similar to the one which ratified the present constitution.

The emphasis should not be on democracy, but rather on creating a workable government, with a culturally relevant national manifesto of state building that could help generate good governance as a prelude to democracy. Elections are too often equated with democracy. In a country like Afghanistan, which lacks democratic traditions and has historically been subjected to traditionalist, authoritarian rule, elections have to be, for some time, simply one of many tools for creating legitimate governments, with a gradual approach to fostering manageable political pluralism, civil-society activities and national reconciliation. Only then could the country secure the foundations for the growth of substantive democracy.

Elections are too often equated with democracy

Reform of the political order should be accompanied by a sound approach to handling Afghanistan's geopolitical complexities. One factor that had historically helped Afghanistan, especially between formal independence in 1919 and the Soviet invasion in 1979, was its pursuance of neutrality. The Afghan–US strategic partnership, signed in August 2005, together with the elevation of neighbouring Pakistan to the status of a major non-NATO ally of the United States, has complicated Afghanistan's regional position. On the one hand, it has insulated Pakistan from the need to make major structural changes in its long-standing predatory behaviour towards Afghanistan. On the other, it has caused deep concern among other regional actors. The Islamic Republic of Iran views a long-term US presence in Afghanistan as a serious threat and remains highly sensitive to any return of Pakistani influence such as existed during the 1996–2001 Taliban regime. Moscow would like to see the end of US and NATO involvement in Afghanistan and Central Asia, which it has historically regarded as vital to its security interests. New Delhi is keen to see Pakistan's regional ambitions curtailed and quietly shares Moscow's opposition to any situation in Afghanistan that could benefit Pakistan. Meanwhile, despite cosy relations with Pakistan and historical differences with India, China shares a common interest with Tehran, Moscow and New Delhi about America's expanding role in the region.

A formal affirmation of Afghanistan's neutrality is needed to reassure Afghanistan's neighbours that the country will not become a long-term base for the United States and its allies, especially Pakistan, to advance their geostrategic interests. This is not a call for Washington to halt assistance for stabilisation and reconstruction, but rather for the activities to be conducted within the framework of UN Security Council resolutions. There is no need for the Afghan–US strategic partnership, which may have stimulated regional actors to remain predatory in their approach to Afghanistan. Nor has the partnership gone down well with most of the Afghan people. Afghans have traditionally seen their country's policy of neutrality as symbolising their devotion to sovereignty and independence.

The Taliban insurgency, moreover, is dominated by the tribally divided ethnic Pashtuns, who make up about 42% of the Afghan population, and inhabit the provinces along the border with Pakistan, with extensive shared cross-border ethnic ties. Karzai and a majority of his ministers are from this ethnic group. The current conflict is essentially an intra-Pashtun one, with the majority of the non-Pashtun population remaining largely aloof. But many non-Pashtuns have grown as disillusioned as their Pashtun counter-parts with the Karzai government and international forces. As their territories in northern, central and western Afghanistan have been relatively peaceful, many feel they have not been rewarded for their cooperation. One way to assuage their concerns is to concentrate international reconstruction investment and efforts in their areas until the Taliban insurgency in the Pashtun-dominated provinces is contained and the way is opened for reconstruction efforts there. Otherwise there is a serious risk that many non-Pashtuns could, despite their experiences under Taliban rule before 2001, engage in insurgency, benefitting the Taliban by default.

As soon as a non-corrupt, efficient and effective government with which Washington and its allies can closely coordinate is in place, and the majority of citizens enjoy a more peaceful and secure life, the conditions would be right for a realistic timetable for the withdrawal of foreign forces and for a viable settlement to be negotiated with the Taliban from a position of strength. At present, any approach to a negotiated settlement is undermined by the Taliban's demand for the departure of foreign forces as a precondition for peace talks, and the inability of the government and its international supporters to leverage the Taliban to settle for an agreement acceptable to a wide cross-section of Afghanistan's mosaic society. The best way forward would be to reach a settlement with the support of Afghanistan's neighbours, predicated on a US–Iranian rapprochement. Without substantial improvement in governance, however, no matter how much assistance the international community pours into Afghanistan, it is likely to continue to go to waste.

Plan B for Afghanistan?

Julian Lindley-French

The clammy odour of defeat is in the air. Talk of a new realism abounds in most of the troop-contributing nations in Afghanistan, although for most it is merely a prelude to an early exit. Both President George W. Bush and Prime Minister Gordon Brown paid surprise visits recently to American and British troops in Afghanistan during which one talked of 'hopeful gains', whilst the other mourned British soldiers who had died in 'the front-line of terror'. The Taliban continue to make gains across the country, while the writ of the Afghan government is increasingly ignored even in Kabul. The Afghan people, the 'critical ground' of the operation, whilst fearing the return of the Taliban, are daily more convinced that the coalition will

Julian Lindley-French is Professor of Military Operational Science at the Netherlands Defence Academy and Senior Associate Fellow at the United Kingdom Defence Academy. The views expressed herein are those of the author and do not necessarily reflect those of any institution.

fail and daily less prepared to offer their support. For them the coalition is simply becoming irrelevant. The next two to three years will thus be critical in the War of the Afghan Succession, if both Afghans and Western publics are to be convinced that something that looks like success can be achieved. Such a goal requires a new mindset in Western capitals.

The arrival of the Barack Obama administration in Washington offers the best chance of such success, not least because a new regime might just be willing to walk away. Indeed, with the appointment of General David Petraeus at US Central Command (CENTCOM) an opportunity now exists for a thoroughgoing review of strategy and an audit of effect across Afghanistan. Only then can a new definition of success be agreed and a shared level of ambition and unity of effort generated across the coalition. What is needed is a Plan B for Afghanistan.

Plan B

All strategies evolve on contact with reality. The refusal by political leaders of all persuasions to face that reality has given ample grounds for journalists and academics of a more excitable persuasion to paint the picture as unremittingly black, with the word 'failure' bandied around as though it is only a matter of time. Washington is rightly committed to a Petraeus-inspired security and development surge similar to the Iraq surge in 2007 and 2008. A particular focus will be the Pashtun heartlands in the south and east of Afghanistan, from which the Taliban draws much of its support, with a renewed focus on pushing and supporting the Pakistanis to deny the Taliban and their foreign associates the safe havens they enjoy in Pakistan's northwest. For all their irresolution, however, European allies afford the Americans all-important legitimacy and thus any such surge must carry allies with it and emphasise the legitimacy afforded by the United Nations mandate.

Any such surge must carry allies with it

Plan B will thus require a mix of modesty and ambition: modesty in fashioning a plan that is achievable, and ambition in that all Coalition partners commit fully over the next two to three years to achieving what is still the vital minimum for success, a belief on the part of the Afghan people that not only is the coalition committed to success, it is going to stay and do the job.

The 2001 Plan A was always ambitious, given Afghanistan's history. The aim was nothing less than a functioning democratic state in which the central authority was recognised by all and corruption was no greater than that with which any functioning state can cope. Whilst all coalition members mouthed their support, only the Americans seemed to believe in such an end. The result was a sovereignty trap by which the coalition pretended the Afghan government was the sovereign body, even though the government has been as much part of the problem as part of the solution. Moreover, the main motivation of many NATO Allies was apparently to do the minimum commensurate with supporting the United States and just enough to keep the Americans interested in the defence of Europe. Very few of the European Allies ever believed in Plan A and it was only at the 2008 NATO Bucharest Summit that an Afghan-wide strategic campaign plan of a sort was fashioned. Even with such a plan it is evident from the lack of concrete support that many NATO allies still regard Plan A as unachievable. Both Afghans and members of the coalition need a plan that is believable.

So what would Plan B look like? First and foremost would be the preservation of an Afghanistan that poses no threat to its neighbours, its own people or the wider world. Realistically, such a goal would require a much looser confederation between Kabul and the tribes. In such circumstances the coalition would then move more into the business of balancing power among the main tribal groups: Pashtun, Hazaras, Uzbeks and Tajiks. Under Plan B reduced Western troop levels would be offset by an over-watch strategy using air assets and special forces backed up by the reinforcing of what passes for pro-coalition elements in the country. Such a shift in force posture would have the added benefit of rescuing many European militaries from the attrition they are suffering from extended and under-resourced stabilisation and reconstruction operations. This would enable the West to re-constitute its armed forces for other possible emergencies. Presentation of such a strategy would of course be vital, to avoid any sense that a shift represents failure (which it would not). The all-important narrative would be self-explanatory: conditions for reconciliation and outreach have been created by the security effort since November 2001 and are thus reflective of a new situation in-country.

Even Plan B would need to see much more commitment by all parties to effective governance capacity-building, establishment of a basic but robust system of rule of law, and a meaningful Afghan national development strategy (reinforced by much greater effort and coordination to ensure Afghanistan can properly absorb aid and development). The lack of credibility now associated with the effort is partly due to unrealistically optimistic forecasts of the time and effort needed to realise something like a stable Afghanistan. The lack of realism at the outset has been reinforced by the hitherto lamentable coalition performance in stabilisation and reconstruction. Indeed, synergy between the various national-led provincial reconstruction teams (PRTs) is frankly awful and this, more than any other factor, has led the coalition to the brink of self-defeat.

Plan B would also see an extension of a regional process with reconciliation front and centre. Maintaining pressure on the Taliban while seeking a dialogue with those elements who are seeking a return would mark a new approach. Indeed, outside the hardcore leadership in Quetta and Peshawar there is evidence that significant numbers of tier-two and -three Taliban could be enticed if property and other rights were restored. At the very least, Plan B would need to reassure the Afghan people that the return of the Taliban to power is not an option.

A regional strategy

Stabilising Pakistan is a prerequisite for stabilising Afghanistan. Thus a far more coherent strategy towards Pakistan is needed. It would at the very least include a coherent package of economic assistance that might require talking to the Chinese and Russians, and possibly even the Iranians. A failing Pakistan will be unable to sustain a campaign against the insurgents in the largely ungoverned tribal areas that border Afghanistan. Pakistan's state institutions must be reinforced, not undermined by coalition action. The commitment of several Western governments to increase aid to Islamabad is to be welcomed, as is a more tailored approach to counter-insurgency operations and a focus on generating effective police forces, vital in both Afghanistan and Pakistan.

India–Pakistan relations are in many ways key to peace in the region, which will prove difficult after the November 2008 attacks on Mumbai. The struggle in Afghanistan has become linked in the official Pakistani mind with the conflict with India over Kashmir. The appearance of several Indian 'consulates' in the south of Afghanistan has only fuelled Pakistani concerns. For the Pakistanis Afghanistan represents strategic depth in the event of a conflict with India. Plan B will demand new strategic partnerships with both Islamabad and New Delhi aimed a de-conflicting Kashmir and Afghanistan. This will not be easy, as the Pakistanis are deeply suspicious of the new US–India agreement over nuclear technology. Europeans would have an important role to play in helping to craft such a strategy.

* * *

Western leaders must come clean with their publics – whilst their armed forces will not stay in strength in Afghanistan indefinitely, the commitment to a stable Afghanistan is a long-term commitment. The Taliban have not defeated the West, but they are doing a pretty good job convincing the West that it has defeated itself. Some places forgive mediocrity. Afghanistan is no such place.

Afghan Q&A: David Petraeus

Toby Dodge, IISS Consulting Senior Fellow for the Middle East:

I wanted to draw General Petraeus out on the balance of resources and troop numbers between Afghanistan and Iraq, and how he would seek to avoid the dangers of robbing Peter to pay Paul.

Douglas Hurd, Former UK Foreign Secretary:

Almost everyone on the panel has emphasised the need for coordinating the civilian and the defence elements everywhere, but particularly in Afghanistan. A great deal of that is being done, and everyone agrees with it. On the ground, however, based on anecdotal evidence very different things are happening. This is really about the structure inside the participating countries – Britain, France, the United States, and so on – where the people who deliver the aid are quite different from the people who do the fighting. They are different in uniform, their bureaucracies are different, their timescales are different, and their psychology is different. This is all the result of institutions' history, which we all in this room understand. However, in an Afghan village where there has been a fire fight, where there are people who are dead, and where immediately you need a clinic, a school, and maybe a magistrates' court within days or weeks, not months or years, how can we break down inside our own systems the division between the civilian and the military? I know there is work going on in this respect, and I would be interested for General Petraeus or Pierre Lellouche to tell us about that work.

David Petraeus, Commander, US CENTCOM:

For Toby Dodge, we have developed a tentative plan that on the one hand allows us in a flexible and responsible way to draw down forces in Iraq – in fact, you may have noted that we actually went to 14 ground brigade combat teams a month or so early – and we have made some other smaller decisions since then; this is from a height of 20 combat brigade teams during the surge. We have a tentative plan and have briefed it as far as the joint chiefs of staff and the secretary of defense, and we will probably take this forward in the next few weeks. It also seeks to answer the requirements that General McKiernan has established in quite a forthright and open manner, both up

5th IISS Regional Security Summit, The Manama Dialogue, Fifth Plenary Session, 'Changing Regional Security Architecture'. For transcripts of the speech and Q&A session see http://www.iiss.org/conferences/the-iiss-regional-security-summit/manama-dialogue-2008/plenary-sessions-and-speeches/.

Survival | vol. 51 no. 1 | February–March 2009 | pp. 97–98 DOI 10.1080/00396330902749731

his US chain of command and his NATO chain of command, which then levies requests on the US. We have determined tentatively how to generate additional forces for him.

The first brigade is already flowing into Afghanistan, having been off-ramped from Iraq. That decision was made, and more recently some engineers were diverted, and there will be other such cases, based on progress in Iraq and based on the enormous increase in the capacity of the Iraqi security forces. The Iraqi army, national police, and special-operations forces just in terms of combat forces alone now have 209 active combat battalions, and over 600,000 total Iraqi security-force members, including police, border police, and so forth. This is quite substantial growth, but the growth is not just in numbers; it is in capability as well. There is still much to be done on the institutional side, with logistics, enablers, and so forth, and those are areas in which we and coalition partners will continue to help.

Minister Hurd asked also about the divide between the civil and military sides, and again Secretary Gates made a critical point on this about how it has to be better joined together in Afghanistan, and not just between coalition partners, the UN, NGOs and IGOs, but also with our Afghan partners. In this regard, a small but very meaningful initiative that was recently approved after being coordinated with Dr Rassoul and the ministers of interior and defence and others, and then approved by President Karzai: a community engagement programme. This will be joined up at local levels. It engages all the members of the community and multiple tribes, for whom there are certain requirements to meet, at which time they then get some resources and they can help contribute to local security so that you have a bottom-up as well as a top-down effort, which we learned in Iraq is hugely important. We look forward to seeing that operationalised in Afghanistan.

Certainly various countries around Afghanistan share some common interests. To take Iran again, Iran certainly does not want to see a Sunni ultra-fundamentalist extremist organisation running their neighbour to the east ... Having said that, I will leave it up to political leaders to determine how to build on shared objectives and perhaps bridge the divide where there are some other very serious differences. That is when it is a good time to be a soldier and perhaps leave the rest to the statesman. Soldiers learn how to go around minefields instead of through them, even at IISS conferences.

Afghan Diary

Rodric Braithwaite

Editor's note

In early September 2008 Rodric Braithwaite travelled to Afghanistan to 'see some of the places (Herat, Pandsher, Salang), talk to some Afghans, and check on the stories my Russian contacts (who go back there all the time) tell me about the positive attitude the Afghans now have towards the Russians'. He kept notes as a record for his own use, and to send back to his family on a daily basis to reassure them that he hadn't been kidnapped. He kindly shared them with *Survival*, and we present a version abridged to fit the available space. Braithwaite did not meet the kind of people that official visitors to Afghanistan usually meet. He drew a number of conclusions from his conversations:

> First, the common belief among Afghans that they were better off under the Russians, and that the last Communist President Najibullah was a better leader than Karzai, may not reflect historical reality; but it bodes ill for the coalition's overall aims. Secondly, the Russian effort in Afghanistan was frustrated above all by their inability to control the frontier with Pakistan: a task which may well turn out to be even harder for the coalition. Thirdly, although the Russians were not defeated militarily in Afghanistan, any more than the Americans were in Vietnam, both the Russians and the Americans failed entirely to achieve their political objectives. The British, by contrast, who ended their nineteenth-century Afghan wars with military victory, sensibly settled for their minimum objective: a monopoly of Afghan foreign policy, which lasted for 80 years. There are no such simple

Rodric Braithwaite, whose diplomatic career included postings in Jakarta, Warsaw, Rome, Brussels and Washington, was British Ambassador in Moscow from 1988 to 1992. He has written two books on Russian affairs and is preparing a third, *Afghantsy*, about the Russians in Afghanistan.

Survival | vol. 51 no 1 | February–March 2009 | pp. 99–118 DOI 10 1080/00396330902749749

options available today. Finally, there is no certainty that the proposed surge in coalition forces will stabilise the country: the analogy with Iraq is just as false as the earlier analogy between the prospects for democracy in Iraq and what happened in Germany and Japan after 1945.

Afghan diary

4–5 September

The KamAir plane was a refurbished Boeing 737, smartly painted in blue and white. The stewardesses were young and pretty, and wore no headscarves. Once we started to come down towards Kabul the mountains loomed barren and light brown, with occasional patches of green in the valleys. The airport was like the airports I knew in Indonesia and the Soviet Union with one or two modern additions, which did not include the luggage-reclaim belt. Its chief defect was that my luggage wasn't on it. I had been told both by BA in London and KamAir in Dubai that there would be no problem, so I had stifled my original instinct to retrieve it in Dubai and check it in again for Kabul. Not to worry, everyone says. It happens all the time and the luggage will turn up tomorrow.

As they went through the airport, the stewardesses were already wearing their headscarves. However, the young American blonde at the exit was carrying a large pistol and certainly no scarf.

The drive in from Kabul airport is dreary – grey dusty ruins interspersed with containers and what look like barracks. No sign of the military. The Turquoise Mountain 'fort' where I am staying is exactly that. It has an ancient wooden gate with huge mediaeval locks on it, and a man roaming around with a Kalashnikov. I asked if he knew how to use it. They said that he did now, because he had just been on a course. The grounds of the fort are now devoted to craft workshops and lecture rooms where they train the students in revived Afghan crafts – wood carving, sculpture, ceramics, calligraphy.

The chief executive, Hedvig Alexander, a Danish woman who was a monitor in Abkhazia during her military service, is married to a former Canadian ambassador to Moscow who is now the chief UN representative here, and has been here six years. She gave me a security briefing – standard commonsense stuff. They are not even allowed to walk in the rather picturesque area around the fort. This is not only because of a rather small risk, but also to avoid drawing too much attention to the fort. However, Hedvig admits that she herself walks around the town without worrying. But she says that the international organisations go into 'White Mode' when there is a heightened threat: they lock down and no one goes outside the fortified compound. That presumably explains why I see so few UN or ISAF vehicles around the town, though there are quite a lot of Afghan police (very corrupt) and the new gendarmerie (rather better).

Saturday 6 September

Zia, one of the Turquoise Mountain drivers, who is a Tajik (and a Sunni) from Pandsher, charming and arrogantly handsome, drives me round Kabul in the morning. I plan to see the Dar-ul Aman palace; the Tadzh Bek palace, stormed by the Russians on the night of 27 December 1979 when they killed Amin, the murderous Communist president; the Bala Hissar castle, next to which the ill-fated 'Army of the Indus' camped before it marched to destruction in 1841; one of the Microrayons built by the Russians for their numerous staff during the war; the Pul-i Charki prison, built by East Germans under the Communists' predecessors.

The town is just what I like – dusty, crowded, the streets lined with open-fronted one-room shops which do everything. Cars, carts, people, men with wonderful faces, veiled women with no face (though not all are veiled), schoolgirls in black cotton jackets and long flapping shirts with white headscarves (two of them are working one of the simple but effective stainless-steel water pumps which I suppose are the product of some uncharacteristically successful aid programme). Men are trading even on the tiny division between the carriageways. We pass one street which is nothing but tinsmiths (there is a pile of tin wheelbarrows outside one of them), and another where they only sell computers and office stationery. There are fruit and vegetable shops and people selling fruit off barrows; meat shops with what look suspiciously like pork carcasses hanging among the flies (presumably it isn't pork in this country). There were a few animals by the roadside, including some multicoloured sheep and goats: presumably this was for identification rather than decoration. The most modern buildings I see are large cubes of coloured glass – four-storey wedding palaces with names like 'Paris Kabul'. Some of the little shops sell white wedding dresses. Almost all of the shops have elaborate painted signs and English wording to say what they do. One beauty shop is called 'Sham and co', well named to judge by the inept picture of a woman advertising their services. Presumably most people can't read the English (or even Dari). Not all the English is perfect: the sign on the traffic-police vehicles reads 'TARAFIC POLICE'. There are big banks with the sort of dodgy names you get in Russia, but also tiny one-room banks (Islamic?). There are a good many police patrols, and we are stopped several times to show our documents. The American Embassy looks open to any smart lad with a rocket launcher. Just outside it there is a huge monument to Masud, the anti-Soviet resistance leader, and now a hero of the Russians as well. The road outside the Indian Embassy is still closed after the recent bomb attack.

The American Embassy looks open to any smart lad with a rocket launcher

The Dar-ul Aman palace, the museum and the Tadzh Bek palace are all some way out of town. The area around is very arid, with the mountains looming above. The Dar-ul Aman palace is very badly damaged, but the walls are mostly standing, and it could presumably be repaired if anyone could be bothered. The Tadzh Bek palace is in much the same state. They seemed to have put a temporary tin roof on it to prevent it deteriorating further. There is an ISAF/Afghan army base close by, presumably where Amin's protection brigade was deployed. The palace is on the top of a hill, and approaches are completely without cover. Storming it was a very tough military operation, and the Russians did well, whatever you think of their motives.

I got into conversation with four young men who were hanging about the Babur Garden. One was called Yama. He was a Pashtun, originally from one of the southern provinces. But his friend was a Tajik from Pandsher, and he emphasised that it made no difference – Pashtuns and Tajiks got on very well with one another. He had studied English, and for a while had worked as a translator for ISAF. But then they wanted to post him to Helmand, and his family had refused to allow him to go. He had been unable to find an alternative job. Things were better under the Russians: everyone had work then – though of course he can't have had any personal experience, he must be reflecting what his family and others are thinking. He was discreetly dismissive of Karzai, who he said was a puppet of the foreigners. He did not pick up my reference to Shah Shujah, the puppet imposed by the British in 1838, to whom Karzai is usually compared – perhaps he had never heard of him, or perhaps I mispronounced the name.

The Bala Hissar, the old fortress near the Dar-ul Aman, is pretty ruined, and I don't think you can get to it because of the mines. But it is high up, and the walls are impressive even now. The road round the bottom is lined with timber yards. The timber apparently comes mostly from North Afghanistan, and must be barely economical to transport.

We then went on to see the 'Microrayons' – five-storey blocks of flats, Khrushchev-style, but with lots of greenery and flowers – quite pleasant. The post-Taliban middle classes live there now: it is the Kabul equivalent of Kensington and Chelsea. One little boy with a kite, who spoke English which he had learned at the elite Habibi School, wouldn't let me take his picture; but three old men did, and one invited me in for tea, but time was pressing by then. Unfortunately both here and at the Bala Hissar the camera had got switched to 'movie' so the pictures didn't come out.

I then went to Shah's bookshop (he is the eponymous bookseller). It too is tiny, a couple of rooms stuffed with the most interesting things. He himself speaks excellent English, intelligent, knowledgeable, and full of charm. I spent too much money there. He too thinks that people lived better when the Russians were here – and he was here at the time. The [Communist] government tried to do something for the ordinary people; and now they see things getting steadily worse. The Russians were not so standoffish as the foreigners are today, who barricade themselves behind high walls and have no interest in Afghanistan itself. He is going to the Frankfurt Book Fair and on to London. He promised to get in touch when he arrives, but I'd be surprised if he does.

Sunday 7 September

The muezzin starts at 4 am, but is very brief.

Although this seemed like a busy day, it wasn't really: I had one appointment, lunch with Sherard Cowper Coles. The other guests were the Russian Ambassador and Christopher Alexander, the Canadian head of the UN mission.

Zamir Kabulov, the highly intelligent and articulate Russian Ambassador, was born in Andizhan in Uzbekistan, and is half Uzbek and half Tajik. He translated when Gorbachev told Khomeini's representative that Islam was more important than Marxism (or words to that effect – apparently it was a famous occasion). He was here for much of the Soviet war.

He expressed polite *Schadenfreude* at the mess the West has now got itself into in Afghanistan. He said that he had spent four months negotiating with the Taliban for the return of the remains of the Russian crew of a downed Mi-8 helicopter working for ISAF: the Taliban were willing to negotiate once they knew the crew were Russian, not Americans. There are two large Russian companies operating the ex-Soviet helicopters here: they stand up much better to local conditions. They are used to supply the ISAF and American forces in the field, and some of the pilots were doing the same thing during the Soviet war. It was a Russian heavy-lift aircraft which brought in the turbine parts that were recently taken up to the Helmand dam by the British Army with so much fanfare: a pointless exercise, since they now have to bring up the cement on which to build the base for the turbines, and if that succeeds they have to build and protect power lines through Taliban territory.

Kabulov said he believed about a hundred Soviets – advisers and their families – were killed in March 1979 during the anti-Communist rising in Herat. The Russians then refused to send troops: it took them another nine months to make up their minds. The Russian military I talk to try to play down the numbers, but the incident did produce a huge shock inside the Soviet political machine – not surprisingly.

All three spoke about the growing nostalgia for the Soviet period. Kabulov agreed when I remarked that your last enemy always seems less bad than your present enemy. There was also a growing feeling that Najibullah had been one of the best of Afghanistan's recent rulers: more popular than Daud, and on a par with Zahir Shah. His policy of national reconciliation had been genuinely intended. If he had survived, Afghanistan would not be in its present mess. This gave the opportunity for a general discussion on Pakistan's role. All thought that *The Bear Trap* (by one of the ISI generals supporting the mujahadeen from Pakistan) was even more relevant today as a guide to Pakistani thinking than it had been when it was written.

We discussed why the Russians have this intense attachment to Afghanistan. I ventured my idea that there were similarities between that and their fascination with the Caucasus. Kabulov was sceptical and Alexander said that the experience in Afghanistan had not produced any comparable artistic and literary response from the Russians.

Kabulov says that Daud was killed in the crossfire during the 1978 coup: Sherard had been to the palace to see Karzai today, and had been up the stairway where it had happened. But the women and children of his family were herded into a separate room and shot in cold blood. Daud's body was then strung up in public – just as Najibullah's was ten years later. They have recently discovered the remains.

They said that the Americans have guard/observation posts in the high mountains like the Russian *zastavas*. And they leave the men there for a similar length of time: up to 16 months. I find it hard to believe that it is possible to do that with soldiers from democratic countries.

Sherard's next appointment was with the newly arrived US commander, to show him pictures of the children killed during a US strike at Shindand a few days ago (last Friday, apparently). Most of the 90 dead were women and children. What seems to have happened is that a group of US special forces got into trouble. They called in an air strike on the basis of intelligence supplied, so the story goes, by the Afghans. The intelligence was faulty – perhaps, conspiracy theorists think, deliberately so, in order further to discredit the Americans. When Karzai went to pay his condolences, one of the local men said: 'We have no more women to look after us or children to kiss'. The US army is still trying to maintain that they were all shot by the Taliban. The pictures seem to demonstrate that this could not have been so.

Monday 8 September

The day began with a security message, apparently put out by the Afghan security authorities over a radio link to NGOs (and presumably others). There was, the message said, a risk of demonstrations today: people should avoid the Babur Gardens. People are celebrating the anniversary of Masud's death, and there are enough people around who didn't like Masud for possible trouble. Helicopters flop overhead, and the traffic seems more congested than usual, but apart from that it looks like an ordinary day. The actual anniversary of Masud's assassination is tomorrow, 9 September, which is a national holiday – something the Pashtuns resent, since they think that thanks to Masud the Tajiks still have too much power.

People look back on the Russian period with nostalgia

I first called on Ismael Saadat, the representative of the Afghan service of the BBC. The BBC office remained open throughout the Taliban era, and he had to attend the executions in the stadium. He says the scene in *Osama* of the execution of a foreign man and woman never happened. He confirms that people look back on the Russian period with nostalgia. Everyone had a job, officials got food parcels every month, there was always something to buy in the shops: it is like the nostalgia Russians feel for the Brezhnev era. The cult of Najibullah is also growing. His speeches calling for national reconciliation and warning that there would be civil war if he was replaced are popular, not least because his dire predictions became true: they are being distributed on DVDs, of which he gave me one. People were deeply shocked when he was butchered by the Taliban.

But because the chaos of the civil war after the Russians left was so dreadful, people were at first glad that the Taliban imposed order. Even Ismael's sister (in law?) looked forward to their coming, though he warned her what it would be like for professional women like her. Better that, and the possibility of walking safely on the streets, she answered.

He agreed with the general view that Karzai was too weak to solve Afghanistan's problems. The economy was at a standstill: the Russians really had provided the elements of industry, whereas most of the aid provided these days simply ended up in the wrong pockets in the wrong countries.

After lunch I went to meet Andrei Greshnov, a good-looking, fair-haired man inclining to flesh. He was here as an interpreter with the Afghan army before the Soviet invasion, and then became a journalist. He is now here with RIA-Novosti.

We met in the Intercontinental Hotel, perched up on one of the hills on the edge of Kabul. The security to get in was quite heavy – slaloms, men with AKs, flimsy-looking lifting barriers, and the usual potholes which, though not designed for the purpose, would certainly slow up a suicide bomber in a vehicle. It all slightly fell apart when you got to the entrance to the hotel, presided over by a doorman in Afghan fancy dress with a notice above him saying 'No Weapons Allowed'. The foyer was completely deserted, and the receptionist seemed to think that I should not be there. Why Greshnov chose this bizarre place to meet I forgot to ask (he later e-mailed to say he had tried to fix the meeting in an Afghan restaurant, but had been brusquely refused by the proprietor because of Ramadan).

Greshnov served as an interpreter with one of the battalions of the 4th Tank Brigade, which spearheaded the bloody Communist coup against Daud. The leaders of the brigade went on to become politicians (some of them still are), but the brigade was broken up and its battalions dispersed. Greshnov served with the battalion in Kandahar.

Greshnov believes that the Russians are and were more popular than the Americans because they too came from poor backgrounds, they didn't give themselves imperial airs, and they didn't dress up in a menacing way. Herat was comparatively unscathed during the war, even though the front line ran right through the city. He had been there at the time. One side of the street was held by the mujahadeen, the other by the Afghan army. But he had been able to walk down the street unharmed, because he was not carrying a weapon. The Russian soldiers had gone jogging along the highway outside Herat, but they were never shot at.

Greshnov said that the Russians made very heavy use of airpower and artillery to support their operations. They had bombed Kandahar on two occasions: one time with aircraft from Tashkent, another with aircraft from

Shindand. Each time – ironically – they had hit the entirely unmilitary area around Pushkin Square. When the Americans had bombed Kandahar, they had also hit Pushkin Square.

Greshnov was adamant that the Russians had nothing to do with the overthrow of Daud. They had worked well enough with Daud for many years, and the coup came as a complete shock to them. But of course they had no choice but to recognise the new regime. I asked him about casualties among the Soviet advisers during the Herat rising. He said there were only two. I told him that Kabulov had said there were a hundred, and that Kosygin had said there were 24 Soviet advisers in Herat at the time. He said there was perhaps a problem of definition, and that the lower figure referred to military advisers only. The worst case he knew of occurred a year or two later, when a busload of Soviet construction workers had been hijacked up in the north. The *spetsnaz* had liberated them, but one of the hostages had been killed in the crossfire – he was the brother of the *spetsnaz* commander.

In the afternoon I was taken by Hilary's friend Nasir Saberi to his home in West Kabul. He lives in the same house that he has had since 1970, though when the Russians arrived he left for London, where he worked for a construction-engineering consultancy. They used to have parties there like the ones in the film of *The Kite Runner*: smartly dressed women, dancing, alcohol. Their Muslim neighbours had not complained. In those days everyone was much more tolerant. After 2001 he was given a job by Karzai, and was responsible for much of the planning and organisation of reconstruction in the ruined city. Karzai was weak. But even Afghanistan's greatest rulers, like Abdur Rahman, had been in the pay of foreigners. Now the Afghans had lost the pride which had always been their distinguishing characteristic.

Saberi had a grand face: clean shaven, but with the right beard it would have been one of those craggy Afghan faces that I would love to photograph, but can't. His brother, a socialist, disappeared during the Communist time, and his family never heard of him again. He is trying to set up something on the lines of the Russian Memorial (my analogy, not his) to discover what happened to the thousands who disappeared. Even if the Russians were not directly involved in the coup (which he did not believe) they bore the moral responsibility for what had happened. And they had waged war with exceptional brutality, wiping out whole villages for no good reason. There is no reason to doubt what he says – but he has the Afghan habit of quoting figures which are clearly grossly exaggerated.

Tuesday 9 September

Just after 8 o'clock I leave for Salang and Pandsher with Zia. On the way out of town we pick up Aziz, who like Zia is from Pandsher.

The roads are generally good all the way, although they break down into potholes from time to time. For the first ten miles out of Kabul it is a double carriageway. Along the road there are the usual endless *dukans*, interspersed with tasteless villas, the product here, as in Russia, of corruption. (I ask Peter Joulwan later in the evening whether the same is true of the excessive number of petrol stations. He says no, they are the result of personal privileges handed out by Karzai and others. I'm not quite sure what the difference is.) The women are all in burkas, and the fields are poorly cared for.

By 10.30 we reach the top of the pass and the entrance to the Salang tunnel. The road is not nearly as precipitous as I expected: I felt no vertigo at any point. I had got a different impression from the Russian accounts I have read. Perhaps the road has been re-engineered, though there is no sign of that. Or perhaps the feeling of dread you get in the Russians' accounts reflects natural anxiety that they might at any moment be attacked.

You approach the tunnel through a series of galleries. The tunnel is shorter than I thought – only three kilometres. But it is indeed scary. There is only minimal lighting, no functioning ventilation, and every now and again some really serious potholes. There isn't too much traffic, and we can just about squeeze past the oncoming lorries. I imagine the awful scenes during the war when one lorry broke down, three convoys were jammed together, and over a hundred Afghans and Russians died of carbon-monoxide poisoning because it was winter and the drivers had kept their engines running to keep warm.

We stop for a bit once through the tunnel, and I see my first wrecked Soviet tanks. We wander about and breathe the fresh air – not nearly as cold as I'd expected – and then turn back through the tunnel again. On the other side I notice, perched on top of some rocks, a little cairn. Zia says it was probably put there by the Soviets as some kind of memorial.

Back down to the turning for Pandsher. After one more little town, we enter the gorge. There is another checkpoint there: but it is a Pandsheri, not a government checkpoint: they want to keep track of foreigners entering or leaving the valley. The gorge really is very narrow, with the fast Pandsher River running below: I find it hard to imagine how the Russians managed to get their huge columns through when they were mounting their offensives here. The sheer logistics must have been almost unmanageable, even without the risk of enemy action.

The wreckage of war is being reused everywhere

From then on the valley broadens out and becomes just as beautiful, green and fertile as its reputation. The fields are well cultivated – maize and corn, by the look of it, and animals as well. There are bits and pieces of tank everywhere: one sprocket wheel has been turned into a sort of flower pot. The wreckage of war is being reused everywhere: it reminds me of the way the farmers on the Somme reuse the barbed-wire pickets. You can see the traces of many ruined villages. But they seem less offensive than elsewhere – the mud bricks simply melt back into the earth they came from.

We drive on to the Masud mausoleum. It is perched on a piece of highland, dominating the valley both up and down. The final shrine is still under construction, but you can see that it will be a tasteless piece of work. The great man's tomb lies under what will be a marble canopy about 20 metres high. Zia and Aziz pray at the tomb. There is no prohibition on photography so I take a few, including some of a group of boys who are larking around some Soviet tanks, and very much want me to photograph them. Aziz tells me he served with Masud during the war. He knew Bystrov, the Russian who converted to Islam and became one of Masud's bodyguards (Peter Joulwan tells me later that Bystrov is well known and respected among the Afghans, rather contrary to what I was told by Nekrasov).

We drive on up the valley and stop. It is Aziz's village. It was flattened during the war, and has been wholly rebuilt. Aziz shows me where the bombs fell. He says: 'That is my house', and asks if I would like to visit it. Of course, I say, and we walk up through his cultivated plots, meet his five small children (whom he kisses passionately – I am immediately reminded of the fathers in Shindand), his donkey, his cows and calves. The house is fairly small, but looks big enough for their needs. I am ushered into the reception room, quite small, with a lovely view of the valley, and cushions round the walls. Far beyond my expectation, Aziz also introduces his wife. She is quiet, intelligent looking, with a headscarf, but politely communicative, except of course that we have no common language. They treat one another with mutual respect. Aziz insists on giving me some of his grapes, and tells me to eat. When I protest that it is Ramadan, he says, 'not for you', and insists. I eat, but I do feel guilty. I keep on realising that I am doing the wrong things – eating with my left hand, showing the soles of my feet (at least I remembered to take my shoes off), but no one seems to mind, and the children are roaring with laughter and turning somersaults. He gives me a large bunch of grapes to take home, and we take our leave: a long, tiring and dusty day, but just what I had wanted. Poor Zia and Aziz are looking forward to breaking their fast: they have eaten and drunk nothing all day, but at 18.15 life will start up again, and they chatter cheerfully all the way back to Kabul.

In the evening I went to meet James Macleod-Hatch at the Gandamack Lodge. This was set up by a British expatriate, Peter Joulwan, who was here as a journalist with the mujahadeen in the 1980s. The security is amazing. There is no sign on the street, just a man lounging about: I assumed that the taxi had brought me to the wrong place. Once they have identified you, you have to go through a series of air locks before you get to the place itself. Even so there is no indication of where the main entrance is. All these measures were introduced after the bombing of the Serena Hotel (Joulwan says that is the very hotel where Dubbs, the American Ambassador, was shot: there are said to be bullet holes still in the walls, but no memorial plaque).

James tells me that there were incoming rockets the previous day – one of the ISAF headquarters and the Ministry of Defence: no damage. He works for ISAF/NATO doing PsyOps. Nothing he says makes me less sceptical of the utility of that stuff. He himself lives in his own hired house, off base. I am a bit surprised: the ISAF security for civilian employees is less stringent than it is at TMF.

When Joulwan arrives he is in local dress. He has converted to Islam, and has an Afghan wife who is running for parliament and is currently in China on a business mission. As a freelance journalist he covered the Afghan war from Pakistan and in the country, but also spent a certain amount of time in Moscow. We had a detailed conversation about Afghan history and politics, and he confirmed that Najib is increasingly the flavour of the month. The better feeling towards the Russians is mainly a matter of nostalgia: people believe the story who are too young to remember. He stuck up for Masud when I relayed some of the criticism. Masud's men had committed atrocities in Kabul during the civil war, but far fewer than Hekmatyar and Sayyaf, and there is no evidence that he gave any encouragement. But he would never have become a major national statesman. He was a brilliant commander, knew how to harness the media, and kept his independence both from the Americans and the ISI, which was more than the other field commanders did. And he always kept his eye on the main aim: ridding the country of foreigners. The others spent a great deal of time feathering their nests and fighting one another.

Wednesday 10 September

Thudding noises from the city at about 7.15 am.

I fly to Herat on the UN plane. They have a separate terminal with security checks every five yards. But the soldiers at the first checkpoint search the car, give us body searches, but don't bother to look into our baggage. The UN plane on which I am flying is a twin turboprop Beechcraft 1300D, with about a dozen seats. The mountains below are burned a uniform brown. As we come in to land, you can see how Herat is entirely surrounded by flat desert, with some mountains in the distance.

I am met at the airport by Abdullah, his friend, and their decrepit Toyota. They were found for me by Turquoise Mountain's local fixer. The driver teaches history of culture, but speaks no English. The young lad who is apparently there to interpret can't understand what I say, and has difficulty in expressing any thoughts clearly. The car works rather well.

We go to the Marco Polo hotel. Manja rings to say my UN flight back to Kabul tomorrow has been cancelled. I quickly call O, the contact I was given in Kabul. He works for ISAF on aid projects. Many members of his family were wiped out by the Communists in January 1979. Afghan soldiers came to their houses and politely asked them to report to the police station. The soldiers were backed up by tanks (with, so O asserted, Russian crews – though how you tell the nationality of the men inside a tank I don't know). The men disappeared, and were never heard of again. O had heard a story that they were taken to Russia and were still alive.

Later, when he was praising Najibullah, I asked how he could talk like that about the man who had killed so many of his relatives. It was God's will that they should die at that time, he replied. I don't understand how this kind of fatalism can be reconciled with the Afghan willingness to fight foreigners or exact revenge.

He tells me that the Taliban were not as bad as they have been painted. They respected women, were genuinely religious, and enforced sharia law effectively. That is what the Afghan people want: but the foreigners won't listen when they are told that.

Meanwhile I agree with Abdullah and co. that they will take me round the sights.

Back at the hotel I pay off the boys and O turns up with the ticket, two books (in Dari – but someone in London will tell me if there is anything of substance in them), and a splendid bearded figure in turban and jacket called Ahmad. He joined the jihad in the first year (1980), became one of Ismael Khan's chief lieutenants, fought with him against the Taliban and fled to Iran when Khan was defeated and the Taliban conquered Herat province, returned with Khan to fight the Taliban in the north, was captured with him but put in a different prison. Khan escaped but Ahmad was in a Taliban prison for five years: he was the last man out of the prison when the Taliban were finally driven off. Thereafter he commanded what he called Security Group No. 43, responsible for the security of Herat city.

The rising in Herat in March 1979, he said, was the consequence of the stupidity of the Taraki government. They failed to realise that it was politically impossible to push through their radical reform programme because it was entirely alien to the culture and traditions of the Afghan people. People were particularly angry about the Communists' attacks on the dignity of

Afghan women. The Herat rising started, in effect, four days before 15 March, in a small village outside the town. The local Communist authorities had tried to implement the edict on female education by taking the local girls to school by force. The villagers had reacted by killing the Communists, and for good measure killing the girls as well. The news spread like wildfire around the other villages, and early on 15 March the peasants marched on the city. The worse clash came in the Char So bazaar: a tank and 30 soldiers were stationed at the crossroads. They killed a thousand people before they were killed themselves. 2,500 more people were killed by two Russian tanks – he thinks.

He thinks that 100 Russians were killed. I asked about the women and children. He says most of the Russians were here without their families. He did not witness the murder of any families. But if they had been found they would certainly have been killed.

25,000 people were killed by government forces in the first three days of the rising. Not all the Afghan army forces in Herat went over to the rebels, and reinforcements were brought in from Kandahar. Bombs fell on the city at the rate of one every two or three seconds for 48 hours. This is a good example of how difficult it is to get reliable information in Afghanistan, and how careful you have to be of 'eyewitness accounts'. Even Hamburg was not subjected to this degree of bombardment by the RAF. There is a persistent story, which still appears in Western accounts, that the Russians carpet-bombed Herat in revenge for the massacre of their advisers, and that they killed 20,000 people. There is no supporting evidence in the documents. The Russians turned down Afghan-government requests to send troops and aircraft to put down the rising, and it was at least six months before they began to decide that they would have to send in their troops. Herat was bombed by a small number of Afghan air-force bombers of Soviet manufacture when the government retook the town. Estimates of the death toll during the Herat rising range from 5,000 to 20,000. If correct, that probably includes the dead on both sides, including those who were massacred by the Afghan Communists both before and after the rising. Estimates for the total number of Afghan deaths during the Soviet war range from 600,000 to 1.5 million: but of course no one knows the real figure. Nevertheless the myths live on.

Ahmad then nips out for a few minutes to pray.

On returning he says that though he fought against Najibullah, he prefers him to Karzai, who was a liar and a player of games. If Najib instead of Karmal had taken over in Kabul when the Russians came in 1979, the country would not be in its present mess.

I asked how he recruited his fighters. He says he and his people moved from village to village, reading the Koran, and calling for jihad. He was with Jemiat-i Islam. They got their weapons from China via Pakistan. To start with they got finance from local rich people. Later the amount of money increased. It was only recently that he discovered that the money had come from the CIA.

He could see no difference between the Russians and the Americans. I said that the Russians were surely more brutal. Not at all, he said. The Russians were strong and brave. They fought man-to-man on the ground, and they used their weapons only when their enemy was armed. They never killed women and children. But the Americans were afraid to fight on the ground, and their bombing was indiscriminate.

He could see no difference between the Russians and the Americans

The next man O took me to was a local worthy, who explained at great length that he was a leading cultural and political figure in Herat, generally known as the 'Encyclopedia of Herat' or 'The Life History of Herat Province'.

I asked him why the Communists pursued such absurd policies after the coup – trying to achieve socialism in five years and shooting anyone who disagreed with them (the regime after 1978 reminds me a bit of Pol Pot, though not quite so extreme). He answered that I would need to understand the whole previous history. Starting with the murder of Nasir Shah and the accession of the 12-year-old Zahir Shah, he took me through 45 years of Afghan history in immense detail. He described the various attempts at democracy up to 1978; the intrigues and coups by which one government had been replaced by another; Daud's success in reinforcing constitutional order, his 'brilliant' policy of buying old weapons from the Russians at inflated prices; and his ability to get Afghanistan respected by its neighbours, including Pakistan.

Daud's coup of 1973 had been facilitated by the two preceding years of drought and floods. He built up the country's infrastructure and industry. But he was too authoritarian in his methods. Meanwhile the Communists had infiltrated into the army and the conditions for a coup were ripening.

By this time he had been speaking for over two hours, almost without a stop, and still hadn't answered my initial question. With difficulty O stemmed the tide by saying that we would have to go to beat the curfew. It was nearly midnight when we left.

Thursday 11 September

A *Hercules* takes off as we are waiting at Herat airport. It fires flares to decoy heat-seeking missiles. I remark on this with surprise: O says that there are some pretty rough bands on the local mountains, and Taliban just down the road.

He says he keeps $2,000 ready in case he has to get his family out of the country quickly – before the roads and airports are closed. He is a very serious, intelligent, devout, patriotic and unhappy Afghan.

In the evening I have dinner with Hedvig and Chris Alexander and Jemima Montague. They live in a small compound in the city with a lovely garden, with heavily armed guards outside the gate. The other guest is a Kosovar, who knew Julian well, and met us very briefly when we visited him in Belgrade in 1997. The discussion was very knowledgeable – Chris, who turns out to be only 40, knows a very great deal about this country, but for professional reasons has to be determinedly optimistic. I asked him why the ubiquitous AK-47 is to be replaced with American guns. He says the American guns are more accurate (though less reliable); that there is clearly an American commercial interest involved; and that the Defence Minister, Wardak, wants the Afghan army to standardise on NATO weapons so that it can operate abroad on peacekeeping missions (perhaps less ludicrous than it sounds, since it would give the army a *raison d'être* apart from just killing other Afghans).

Friday 12 September

Zia – who was flying his fighting kite yesterday and lost – takes me to the airport. It is the most confusing departure procedure I have ever experienced, with body frisks every ten yards, and people insistently offering their services to carry bags, have them strapped up, help them through customs and other barriers. I cannot get rid of all of them, and some happy Kabulis end up rich as I unload my afghanis on them. By contrast the elderly man keeping the departure lounge clean gets me some tea and refuses to accept payment for it. He asks me how many children I have: five. I ask him: none. I show him pictures of my *khanum* and *dochir*. All this with barely a word in common.

The Problem with Nuclear Mind Reading

James M. Acton

The basic technologies that underlie nuclear power were first developed during the Second World War by the United States for purely military purposes. Since then, extensive civilian research and the use of nuclear technologies for peaceful ends has not erased this military heritage. Almost all nuclear technology in use around the world today is 'dual use', able to contribute to the production of fuel for nuclear reactors or the explosive components of nuclear weapons. For this reason, there is serious concern that supposedly peaceful nuclear programmes are being used for, or could become, cover for the development of nuclear weapons.

The inherent ambiguity surrounding almost all nuclear technology complicates the control of nuclear energy. Under the terms of the Nuclear Non-Proliferation Treaty (NPT), non-nuclear-weapons states must declare all their nuclear facilities and activities and permit them to be safeguarded by the International Atomic Energy Agency (IAEA). To date all cases of non-compliance have involved states failing to declare nuclear facilities or activities, rather than diverting safeguarded nuclear material.[1] In the last few years, for instance, clandestine nuclear activities have been uncovered in Egypt,[2] Iran[3] and South Korea (Republic of Korea),[4] and strong evidence of a clandestine reactor in Syria has emerged.[5] These discoveries have prompted debates, of varying intensity, about what, if any, action should be taken in response.

James M. Acton is an associate in the Nonproliferation Program at the Carnegie Endowment for International Peace in Washington DC. Previously, he was a lecturer in the Centre for Science and Security Studies, part of the Department of War Studies at King's College London, where this work was partially undertaken. He is co-author of the recent Adelphi Paper *Abolishing Nuclear Weapons*.

Survival | vol. 51 no. 1 | February–March 2009 | pp. 119–142 DOI 10.1080/00396330902749756

Given the dual-use nature of the technology involved, these debates have focused on intentions: were the clandestine activities for peaceful purposes or were they part of a nuclear-weapons programme? British, French and American diplomats regularly argue for tougher sanctions on Iran, for instance, on the basis that the country has not done enough to demonstrate that its intentions are benign. Other states oppose tougher sanctions, arguing that Iran's intent to proliferate has not been proven.[6] Similarly, the movement to refer South Korea to the United Nations Security Council in 2004 was blunted because most in the international community were content to give the Koreans the benefit of the doubt regarding their intentions.

The focus on intent in these debates is unsurprising. The notion that intent is relevant to enforcement decisions is familiar from domestic law: the reasons for a killing determine whether it constitutes murder or manslaughter, for instance. Extrapolating to international law, it certainly seems logical that a non-compliant state's intentions should be central to the debate over the international community's response.

Moreover, the importance of intent is supported by the standard interpretation of Article II of the NPT. This article contains the injunction on non-nuclear-weapons states 'not to manufacture or otherwise acquire nuclear weapons'.[7] During negotiations, the definition of the term 'manufacture' was contentious,[8] but agreement was obtained on a set of criteria which state that 'facts indicating that the *purpose* of a particular activity was the acquisition of a nuclear explosive device would tend to show non-compliance'.[9] These so-called Foster Criteria suggest that if a state develops enrichment technology (clandestinely or otherwise) with the intention of using it to produce nuclear weapons, it has violated Article II. When its clandestine nuclear activities were uncovered, Iran claimed that they were directed solely towards the production of fuel for nuclear reactors and hence that it had not violated Article II. In a similar vein, South Korea claimed that its clandestine activities were basic research conducted without the knowledge of the government.

Against this background, some may find it surprising that the IAEA has generally not attempted to assess *why* non-compliant states have conducted undeclared activities. Instead, the agency has, in accordance with its legal mandate, almost exclusively concerned itself with documenting *what* states have done. This policy is sensible and inevitable. There are very few activities, materials or equipment unambiguously associated with the manufacture of nuclear weapons, and the chances are slim that the IAEA would detect one that was. Moreover, without recourse to the much more intrusive forms of information-gathering employed by national intelligence agencies, the IAEA has equally little chance of proving why an 'ambiguous' activity was conducted.

There is, therefore, a significant lacuna in the non-proliferation regime: the IAEA Board of Governors and the Security Council place great importance on something – intent – that their investigatory arm does not and could not assess. Thus, the effect of emphasising intent is to cause discussions about enforcement to descend into irresolvable arguments about ambiguous activities, significantly reducing the chances of robust, decisive action. This in turn reduces the credibility and effectiveness of the non-proliferation regime and should prompt the international community to rethink the way it enforces non-proliferation agreements.

Misinterpreting the IAEA

Much of the debate over an appropriate response to Iran's clandestine nuclear activities has centred on whether they were conducted for peaceful purposes. There is a widespread assumption, encouraged by Iran itself, that the task of determining intent falls on the IAEA. For instance, commenting on the IAEA's 22 February 2008 report,[10] the South African ambassador to the Security Council stated, 'those issues that originally gave rise to serious concern ... have now been clarified' and that, as a result, 'there ought to be increased confidence in the peaceful nature of the Iranian nuclear programme'.[11] This interpretation was widespread, but is a significant misrepresentation of the IAEA's findings.[12]

One of the principal issues discussed in the report was a series of experiments to produce and separate polonium-210, a material that can be used in a device to initiate the chain reaction in a nuclear weapon. In the report, the IAEA repeats Iran's claims about its intentions, that is, that the experiments were 'fundamental research aimed at enhancing knowledge'.[13] It also lists the supporting evidence put forward by Iran. However, the report concludes:

> Based on an examination of all information provided by Iran, the Agency concluded that the explanations concerning the *content* and *magnitude* of the polonium-210 experiments were consistent with the Agency's findings and with other information available to it. The Agency considers this question no longer outstanding at this stage.[14]

This statement neither supports nor disputes Iran's claimed intentions. The only parts of Iran's explanation verified by the IAEA were purely technical claims about what the experiments involved and how much polonium was extracted. In other words, the IAEA concerned itself with verifying what Iran did, not why it acted. Similarly, the IAEA did not come to any conclusion about Iran's intentions when it relegated questions of the country's plutonium-extraction experiments,[15] the source of contamination at a technical university,[16] and the country's centrifuge programme[17] to the category of 'matters no longer outstanding'.[18]

The IAEA's policy of not assessing intentions is not limited to Iran. In spite of media reporting to contrary, the IAEA's investigation into South Korea's undeclared nuclear activities did not conclude that 'there wasn't any evidence South Korea was trying to make nuclear weapons',[19] but rather that 'there is no indication that the undeclared experiments have continued'.[20]

The IAEA has no legal basis on which to draw conclusions about intent

Indeed, the IAEA has no legal basis on which to draw conclusions about intent. The IAEA derives its mandate to apply safeguards in all non-nuclear-weapons states from Article III.1 of the NPT. As required by this article, the detailed arrangements for the application of safeguards are set out in a subsidiary agreement, the model Comprehensive Safeguards Agreement (INFCIRC/153). This states that the objective of safeguards is 'the timely detection of diversion of significant quantities of *nuclear material* from peaceful nuclear activities to the manufacture of nuclear weapons or of other nuclear explosive devices or for purposes unknown'.[21] Given the exclusive focus of this article on nuclear material, it is hard to argue that the IAEA has the right to assess intentions. The only possible exception is when the IAEA investigates weaponisation activities (activities connected with the design or fabrication of nuclear weapons). The current IAEA director general, Mohamed ElBaradei, has stated that the agency may investigate such activities where there is 'some nexus linking the activity to nuclear material'.[22] If the IAEA uncovered weaponisation activities (as it has tried to do in Iran) it would also produce proof of the intention to build nuclear weapons. However, proof of intent is a 'by-product'; investigations into weaponisation activities are permitted only because their purpose is to detect undeclared or diverted nuclear material.

Could the IAEA assess intent?

Regardless of whether the agency is currently entitled to assess intent, it could, in theory, be tasked with doing so. There are two ways that proof of a state's intentions could be obtained. (Whether it is really reasonable to expect proof – as opposed to good evidence – is an important issue, but one that will not be discussed any further here since proof seems to have become the operative standard.[23]) Firstly, if activities or equipment *unambiguously* associated with the development of nuclear weapons were uncovered, there could be no uncertainty about a state's intentions. Such evidence is, unfortunately, very hard to obtain.[24] The overwhelming majority of activities and equipment associated with the manufacture of nuclear weapons is dual use.[25] The few exceptions, such as uranium or plutonium hemispheres used in the 'pit' (metallic core) of a nuclear weapon, are easy to conceal. Moreover, a state would be extremely unlikely to manufacture any such objects in a declared nuclear facility that was subject to regular inspections. Although the IAEA does have the authority to visit undeclared facilities,[26] the time taken to negotiate access would give the state ample opportunity to remove any incriminating items. For the IAEA to have any reasonable chance of finding such unambiguous evidence it would need 'any time, anywhere' access (and even then success would be far from guaranteed without good background information to help inspectors know where to look). The IAEA has only ever been granted such intrusive access in Iraq, in 1991 and again in 2002, and that was *after* the Security Council became convinced that Iraq had been actively seeking nuclear weapons.[27]

The difficulties of proving intent are borne out by experience in Iran. There, the IAEA succeeded in uncovering the notorious '15-page document', which describes how uranium can be cast into hemispheres, a process that is unambiguously associated with the manufacture of nuclear weapons.[28] Yet even this, it seems, did not constitute proof of Iran's intentions. In its defence, Iran claimed that the document was given to it, unsolicited, by the Pakistan-based A.Q. Khan network. That this explanation was apparently considered plausible illustrates how difficult *proving* intent really is. More recently, the focus of the IAEA's investigation has shifted to the so-called 'alleged studies', which involve a number of activities unambiguously associated with weaponisation, including the 'symmetrical initiation of a hemispherical high explosive charge'.[29] To date, however, it has not been possible to prove that these activities took place.[30]

Alternatively, the IAEA could determine a state's intentions if it came across evidence proving that a dual-use activity (such as the production of polonium-210) was carried out for a proscribed purpose. National intelligence agencies can plausibly obtain such information by intercepting conversations between senior officials or through well-place agents. However, signals intelligence and human intelligence are not means that the IAEA currently has at its disposal, and it is frankly impossible to envisage states ever giving the agency permission to spy on them in this way. Moreover, it would probably be undesirable for the IAEA to have such powers. The agency relies on the cooperation of states to facilitate inspections, provide information and answer questions. Permitting the IAEA to gather covert intelligence would lead to a much more adversarial relationship between states and the agency, significantly undermining the ability of the latter to do its job effectively.

Although the IAEA cannot use intelligence-gathering techniques, it can request interviews with people associated with nuclear activities. Absent specific authority from the Security Council, however, states are under no legal obligation to accede to such a request.[31] Iran, for instance, has been generally resistant to providing inspectors with access to personnel and has not permitted key individuals to be interviewed.[32] Moreover, interviews can only be effective in determining intent if the agency has access to those at the heart of a suspected nuclear-weapons programme, as they may be the only people who actually know what is going on. The scientists in Iran who conducted the experiments to produce polonium-210, for instance, may not have been told that they were contributing to a nuclear-weapons programme (if indeed they were). Thus, by carefully selecting which scientists and officials the IAEA is allowed to interview, a state could potentially hide its intentions.

Alternatively, a state can undermine the effectiveness of interviews by insisting they take place in the presence of government officials, thus tacitly threatening retaliation if the interviewee gives the 'wrong' answer. This has occurred in Iran.[33] In theory, this problem might be circumvented if the Security Council were to insist that interviews take place abroad and that family members be permitted to accompany interviewees. However, even apart from the reluctance of the Security Council to authorise such drastic measures (it has only ever granted this power once, in Iraq in 2002),[34] it would hardly be a practical solution if hundreds of people were wanted for interviews.

Given all these limitations, the only means left to the IAEA to assess intent is to analyse proliferation drivers, trends in capabilities and public statements (much as national intelligence agencies are often forced to do).[35] Indeed, the agency already does this, but only for its own internal planning; results are not publicised. Moreover, because this kind of analysis cannot yield proof, asking the agency to make such information public could actually be counterproductive. The institutional culture of parts of the IAEA makes the agency extremely reluctant to raise suspicions without proof – good evidence is not enough.[36] Requiring that the IAEA make its intent estimates public could lead it to whitewash states, even if the evidence was much more ambiguous.

Good evidence is not enough

Beyond the IAEA's internal culture and limited inspection authority – both of which could possibly be changed – there are two more fundamental difficulties with assessing intent. Firstly, nuclear-weapons programmes do not consist of a single decision that results in the construction of an atomic bomb some years later.[37] Rather, such programmes always involve a number of decisions to scale activity up or down, and a programme's nature may change gradually over time. For example, from 1958 Sweden adopted a policy of 'deciding not to decide'.[38] Swedish researchers conducted extensive nuclear research while being 'formally forbidden to cross the line into "research explicitly intended to result in the construction of an atom bomb"'.[39] Their work produced knowledge relevant to the manufacture of nuclear weapons and prevented existing knowledge from being lost. Ultimately, after years of ebbing intent, Sweden finally decided to abandon this policy in the early 1970s. The fact that intent actually lies along a continuum and is not constant does not make it impossible to determine, but it does make it much harder than if intent were an unchanging binary variable.

Finally, and most importantly, many states do not have a single centre of power that sets national policy. Instead, often by design, there may be many competing entities vying for influence. Iran, with its supreme leader, president, cabinet, parliament, Guardian Council, Expediency Council, Assembly of Experts and Islamic Revolutionary Guard Corps, provides a classic example. It is entirely possible that some of these organs support the development of nuclear weapons while others do not. A less hypothetical example is again provided by Sweden. By the early 1960s, the ruling Social Democrats (along with the majority of the population) were generally opposed to the acquisition of nuclear weapons, while the military supported it and, with the government's permission, continued to conduct relevant research.[40] Clearly, the whole concept of a state having well-defined intentions is problematic.

Why intent shouldn't matter

The international community's efforts to enforce arms-control agreements can be divided into two different conceptual 'strands'. The 'multinational-bodies' strand involves the work of the Security Council and other international organisations specifically tasked with enforcement (such as the IAEA Board of Governors, which has the important task of referring a state that has violated its nuclear-non-proliferation obligations to the Security Council).[41] The tools available to this strand are primarily punitive. The second strand is the 'state-to-state' strand, which consists of ad hoc efforts made by a single state or a coalition. In the case of Iran, this strand is epitomised by the work of the E3 (the United Kingdom, France, Germany) and, more recently, the E3+3 (the E3 countries plus the United States, Russia and China). Direct bilateral talks between the United States and Iran (were they to occur) would also fall under this strand. In contrast to the multinational-bodies strand, incentives to induce compliance, as well as punishments to compel it, are available. This division is not perfect – it is unclear, for instance, where the Proliferation Security Initiative would fit in – but it is useful for current purposes.

A non-compliant state's perceived intent does – and should continue to play – a key role in the work of the state-to-state strand. It would be perverse to suggest, for instance, that in preparing for negotiations, Washington should ignore its intelligence community's assessment of Iran's intent. Moreover, a package of incentives (such as those offered to Iran in June 2006 and June 2008) can be tailored much more carefully with an understanding of a violator's intentions. Similarly, the United States has helped to prevent proliferation in the past by convincing its allies not to export proliferation-sensitive nuclear technology because it feared the intended recipient's intent.[42]

Within the multinational-bodies strand, the IAEA plays a central role by providing the international community with an independent assessment of a state's actions. However, for all the reasons discussed above, not only does the IAEA not attempt to assess intent at the moment, it is hard to envisage it ever doing so. By taking a non-compliant state's intent into account without independent means to assess it, the Board of Governors and Security Council are undermining their ability to prevent proliferation. There are two distinct reasons for this. Firstly, because proof of intent is needed before meaningful enforcement actions are authorised, potential proliferators may rightly conclude that they will face minimal consequences if caught. This erodes the deterrence value of safeguards. Secondly, when deterrence fails, this same inability to agree upon a course of action severely hinders efforts to prevent a state that is actively seeking nuclear weapons from acquiring them.

Preventing a state that has embarked on a nuclear-weapons programme from succeeding is a difficult and messy business. Successful prevention is not guaranteed. The most effective form of non-proliferation is to dissuade states from choosing to develop nuclear weapons in the first place. For this reason, the deterrence function of safeguards is vital to the success of the non-proliferation regime. The effectiveness of deterrence depends upon a would-be proliferator's perception of two different factors: (1) the probability that the IAEA would detect any non-compliance and (2) the likelihood and severity of the international response. In general, a state will only be deterred if it fears both detection and its consequences.

No matter how good IAEA safeguards are, if proliferators believe they can avoid or severely delay any consequences, whatever deterrence value safeguards have will be severely eroded. One lesson that the next proliferator can draw from the case of Iran is that if its non-compliance is detected, it should play the 'intent card' and argue that its intentions are benign. Given that the IAEA cannot assess intent (and states rarely share the intelligence data on which national estimates of intent are based), this strategy would help mire the Board of Governors and Security Council in an essentially irresolvable debate that would likely prevent decisive action. Deterrence is most effective if clear 'red lines' exist, the crossing of which can be detected relatively unambiguously. Such red lines exist in the form of a state's safeguards agreement and for that reason non-compliance with it ought to be the standard for initiating enforcement actions. The more intent is considered, the less distinct the lines become.

The focus on intent also undermines the ability of the international community to manage a state that has already started down the road of acquiring nuclear weapons. By insisting upon proof of intent before acting meaningfully, it is inevitable that the Security Council will sometimes fail to enact robust sanctions against a proliferator. Indeed, the Board of Governors might even fail to refer the case to the Security Council. (This is not to say that the Board of Governors and the Security Council will always succeed in preventing a would-be proliferator from acquiring nuclear weapons if they do act robustly, but by not doing so they have little chance of success.) Handing the benefit of the doubt to a non-compliant state is foolhardy where non-proliferation is the goal. The alternative – taking action against any state that violates its safeguards agreement, regardless of intent – is hardly ideal, as it may sometimes result in an 'innocent' state being 'punished', but a state that found itself in this position would have only itself to blame – after all, it would have violated an international agreement. Ultimately, this outcome is preferable to allowing proliferators to avoid meaningful consequences.

Giving benefit of the doubt to a non-compliant state is foolhardy

There is a second, more subtle reason why the current focus on intent is corrosive. States do not decide whether to support enforcement actions purely on the basis of evidence presented by the IAEA. Such evidence does matter, but states are also swayed by narrower considerations of self-interest, such as their bilateral relationship with the non-compliant state. States are likely to decide upon enforcement actions on the merits of the case, rather than on the basis of self-interest, if the salient factors are those that can be reliably and accurately assessed by the IAEA. Intent cannot. Thus, if intent occupies a central role in the deliberations of the Board of Governors or Security Council, it becomes easier for member states to prioritise self-interest above non-proliferation.

This has occurred in the case of Iran. The focus on intentions has allowed Russia in particular to downplay the compelling evidence of non-compliance presented by the IAEA. Moscow has frequently argued that the United Nations should not impose or ramp up sanctions in the absence of proof that Iran is seeking nuclear weapons.[43] Whether or not Russia genuinely questions Iran's intentions, it is clearly politically convenient for it to do so. Had the international debate over Iran focused more on its actions, there would have been less political space for the Russians to oppose sanctions. Indeed, since the implementation of Security Council Resolution 1696 (2006), the most persuasive argument for convincing Russia to support subsequent resolutions imposing sanctions has been Iran's failure to abide by the Security Council's demands – an argument about Iran's actions, not its intentions. For instance, Russia's ambassador to the United Nations justified his nation's decision to support Resolution 1803 (2008) by saying, 'the resolution is in fact a political signal to Iran of the need to cooperate with the international community by implementing the decisions of the Board of Governors of the International Atomic Energy Agency (IAEA) and the Security Council'.[44]

The more decisions about enforcement are made on the basis of self-interest rather than the merits of the case, the more a rules-based system like the non-proliferation regime is corroded. If the Security Council or Board of Governors focuses on intent, the inconclusive nature of the evidence can make it easy for one of the United States, Russia or China to oppose enforcement actions when they contradict short-term considerations of self-interest. Given that all three states possess a Security Council veto, and that at least one will almost always have a self-interested reason for opposing enforcement, this is a recipe for preventing the international community from ever acting.

Assessing the severity of non-compliance

A frequent objection to the idea of de-emphasising intent in cases of non-compliance is that this would lead to all safeguards violations being treated equally: while all non-compliant states might receive an automatic 'slap on the wrist', it would become even harder to build a consensus for robust action in the most serious cases.[45] However, as the following tentative framework demonstrates,[46] there are other grounds for assessing the severity of non-compliance:

1. *The quantity and 'strategic significance' of the nuclear material involved.* The actual amount of fissile material clandestinely produced or diverted is one obvious factor in assessing severity, although it is important to note that highly significant research and development can be conducted with very small quantities of material. In addition, although any nuclear material is potentially relevant to a nuclear-weapons programme, some types are more useful than others (highly enriched uranium metal is more significant than

low-enriched ceramic reactor fuel, for instance). For this reason, the 'strategic significance' of the nuclear material involved could be considered.

2. *Capability*. The closer an undeclared programme is to giving a state the capability to manufacture fissile material in militarily significant quantities, the more serious the breach could be considered. The existence of facilities (declared or undeclared) that would permit clandestine experiments to be quickly exploited on a larger scale could be an aggravating factor.
3. *Intensity*. A focused, concerted effort at developing fuel-cycle technology could be considered more serious than a few isolated experiments.
4. *Termination*. A voluntary decision by the state to terminate its non-compliance *before* discovery could be a mitigating factor.
5. *Military dimension*. The involvement of the military could be an important aggravating factor, as could the discovery of experiments with potential relevance to weaponisation.
6. *Concealment*. Active attempts to conceal the non-compliance while it was occurring could be an aggravating factor.
7. *Detection*. The voluntary self-reporting of non-compliance to the IAEA (as Romania did in 1992)[47] could be a mitigating factor.
8. *Cooperation with the IAEA*. Failures to cooperate actively with an IAEA investigation, or even more seriously, active attempts to impede it, could be an aggravating factor.
9. *Compliance with IAEA Board of Governors or Security Council resolutions*. A failure to abide by relevant legally binding resolutions could compound the seriousness of a violation.

These criteria are already routinely assessed by the IAEA when it investigates non-compliance (as its reports on recent investigations make clear, though they are not structured in this form). Some of these criteria undoubtedly correlate with intent, albeit imperfectly. However, unlike intent, they can all be measured objectively by the IAEA within its current mandate.

Table 1: **Comparison of the nuclear-safeguards violations of Iran and South Korea**

Criteria	Iran[1]	South Korea[2]
Quantity and strategic significance of the nuclear material involved	Small amounts of LEU and separated plutonium were produced in bench-scale experiments and pilot-scale facilities: • Centrifuge experiments involved 1.9kg UF_6 feedstock, enriched to probably no more than 1.2%; • Bench-scale AVLIS experiments involved the evaporation of 0.5kg of uranium metal and the collection of milligram quantities of product enriched to an average of 8% and a maximum of 13%; • Pilot-scale AVLIS experiments used 0.5kg uranium metal and resulted in enrichment levels of less than 1%; • Depleted uranium was irradiated to produce 0.1g plutonium, of which milligram quantities were separated. About 0.5t of various uranium-containing compounds were used in a number of undeclared conversion experiments. The import of about 2t of various uranium-containing compounds was undeclared.[3]	Very small amounts of LEU, HEU and partially separated plutonium were produced in bench-scale experiments: • AVLIS experiments produced 200mg of uranium enriched to an average of 10% and a maximum of 77%; • A 'very small' quantity of uranium was enriched via chemical exchange; • Depleted uranium was irradiated to produce 0.7g of plutonium, of which some was partially separated but not converted to metallic form. 154kg of undeclared natural uranium was produced in conversion experiments.
Capability	Very high. Iran's undeclared activities were part of a systematic attempt 'to master an independent nuclear fuel cycle'. In particular, its undeclared centrifuge work was a concerted effort to develop a significant enrichment capability. Even the pilot-scale AVLIS system 'would have been capable of HEU production had the entire package of equipment been delivered'. Iran failed to declare a pilot-scale centrifuge facility (PFEP). It possessed or was building production-scale facilities (UCF and FEP) to take advantage of its undeclared research into and development of centrifuge technology.[4]	Medium. No part of South Korea's undeclared enrichment, irradiation or plutonium-separation experiments progressed beyond laboratory scale. However, like Iran, the country experimented with uranium-metal production (the feed material for AVLIS). The throughput of the conversion facilities is not known.
Intensity	Very high for some aspects of the programme at certain times. Centrifuge work, for instance, was particularly focused and intense after about 1997. AVLIS and conversion research appears to have been fairly focused. Plutonium production and separation research was isolated and sporadic.	AVLIS and conversion work was reasonably focused. Chemical enrichment and plutonium production and separation research were isolated and sporadic.
Termination	Ongoing when detected in 2003.	Terminated voluntarily before discovery.
Military dimension	Yes. Centrifuge components were manufactured in at least one workshop on a Defence Industries Organisation site. Potential weaponisation activities included (1) possession of the uranium-metal document; (2) production of polonium-210; and (3) the 'alleged studies' which involve high-explosive development (including detonator development, hydrodynamics testing and plans for an underground test facility) and designs for a nuclear-armed missile re-entry vehicle.	No.

Table 1: **Comparison of the nuclear-safeguards violations of Iran and South Korea**

Criteria	Iran[1]	South Korea[2]
Concealment	Extensive. Iran went to great lengths to keep its programme secret (for instance, it made use of the nuclear black market to procure centrifuge technology, built underground facilities, and submitted incorrect reports to the IAEA).	Unclear. Seoul twice refused a voluntary 'transparency visit' to the Laser Technology R&D Centre but the government claims it was unaware that nuclear material had been used there. It also issued incorrect reports in the early 1980s about the material used in the irradiation experiments. In contrast, chemical-exchange activities were discovered from published literature.
Detection	The existence of the enrichment facilities at Natanz was publicly revealed by an Iranian resistance organisation. Subsequent IAEA investigations revealed the extent of Iranian non-compliance.	The country voluntarily reported on its AVLIS experiments, but only after the IAEA had become suspicious. The plutonium-separation experiment, certain conversion activities and the chemical-exchange experiment were discovered as a result of IAEA verification activities.
Cooperation with the IAEA	Generally poor, including the use of 'extensive concealment activities' to hinder IAEA investigations (such as the renovation of the Kalaye Electric Factory in an attempt to interfere with environmental sampling and the provision of incorrect and misleading information to inspectors). During the IAEA's five-year investigation Iran has cooperated more actively at points but significant issues remain, including the suspension of Additional Protocol implementation, the suspension of modifications to Iran's Subsidiary Arrangements,[5] and a refusal to enter into meaningful discussions about the alleged studies.	Active.
Compliance with IAEA Board of Governors or Security Council resolutions	Consistent failure to comply with IAEA Board of Governors and Security Council demands. For instance, work on centrifuge installation at the FEP and the construction of the IR-40 reactor at Arak is continuing. Centrifuge operations in both the PFEP and FEP are ongoing (from February 2007 to November 2008, 9,750 kg of UF_6 had been fed into the cascades at the FEP).	Not applicable (no relevant Board of Governors or Security Council resolutions were passed).

Key

AVLIS: Atomic vapour laser isotope separation (a type of laser-enrichment technology)
LEU: Low-enriched uranium
HEU: Highly enriched uranium
FEP: Fuel Enrichment Plant (Natanz, Iran)
PFEP: Pilot Fuel Enrichment Plant (Natanz, Iran)
UCF: Uranium Conversion Facility (Esfahan, Iran)

Notes

1 All information obtained from the IAEA Director General's reports on Iran, available at http://www.iaea.org/NewsCenter/Focus/IaeaIran/index.shtml.

2 All information obtained from IAEA, GOV/2004/84.

3 In addition, Iran had been very slow in declaring some previous imports. Most notably, the import of 531t of U_3O_8 in 1982 was not reported until 1990.

4 The Uranium Conversion Facility was declared to the IAEA in 2000, although the design information Iran provided was later shown to be incomplete. The Fuel Enrichment Plant was not declared to the IAEA until February 2003. Technically, this was not a safeguards violation as Iran's Subsidiary Arrangements only obliged it to declare a new facility 180 days before the introduction of nuclear material. At that time, however, Iran was the only state that had not accepted a modification to its Subsidiary Arrangements that required it to declare a new facility as soon as it took the decision to build one. Iran subsequently accepted the modified subsidiary arrangements on 26 February 2003, but on 29 March 2007 announced it would no longer be bound by them. Iran's failure to declare the Pilot Fuel Enrichment Plant was a safeguards violation under either version of the Subsidiary Arrangements. See Pierre Goldschmidt and George Perkovich, 'Correcting Iran's Nuclear Disinformation', Proliferation Analysis, Carnegie Endowment for International Peace, 27 March 2007, http://www.carnegieendowment.org/npp/publications/index.cfm?fa=view&id=19078.

5 See note 4.

Using this framework, Iranian non-compliance is analysed in Table 1. For the purposes of comparison, a similar analysis is presented for South Korea. It is clear that these cases are far from equal and that there are grounds (other than intent) for judging Iran's activities to be much more serious than South Korea's. As a result, the fear that it would be necessary to treat all cases of non-compliance equivalently would appear to be misplaced.

Another criticism that has been made of this approach is that it is overly 'mechanistic' and would lead to action being taken when it was not merited, as in the case of a state that has made a bureaucratic error in its reporting to the IAEA. However, the IAEA (quite properly) does not consider genuine reporting errors grounds for non-compliance, and such cases are not referred to the Board of Governors or Security Council, so the question of assessing the severity of non-compliance is moot.

Another example of non-compliant behaviour that many believe should not merit an international response because of the absence of malign intent is the failure to meet a deadline for destroying munitions under a disarmament agreement. In particular, both the United States and Russia look likely to miss the destruction deadline set by the Chemical Weapons Convention. Although this article focuses on the Non-Proliferation Treaty and its associated safeguards instruments, this example presents an interesting conceptual challenge. Under the framework outlined above, such a failure would only constitute a comparatively minor violation provided that the state in question had been transparent (criteria 6). However, such a violation *would* deserve censure, albeit of a relatively mild form. For all the reasons discussed above, a rules-based system is eroded if exceptions are continually made. By being willing to accept censure where they fail to meet the terms of an agreement, the United States and Russia would set a precedent that would help to build a consensus for action when other states did likewise.

Is downplaying intent practical?

One important question remains: is a policy of avoiding questions of intent actually practical? Can the international community ever be persuaded to agree upon robust enforcement actions while doubts about intent remain?

An answer to this question is that there is much the United States and its friends could do *unilaterally* to reduce the emphasis on intent. The steps outlined below would certainly not remove intent entirely from international deliberations, but they might gradually move it into the background and create an environment in which meaningful enforcement actions could be agreed upon in the presence of doubts about a non-compliant state's intentions.

Firstly, the United States and its allies would almost certainly bring about a reduced focus on intent if they stopped using it as an argument themselves. As discussed above, Russia became somewhat more receptive to tightening sanctions against Iran when the argument was made in terms of Iran's refusal to abide by Security Council resolutions rather than in terms of its intent. If the case against future non-compliant states is put in terms of proven safeguards violations (and non-compliance with Board of Governors and Security Council resolutions) it will probably result in a tougher response.

Secondly, the deterrence value of safeguards would be strengthened if the Security Council were to adopt a generic, legally binding resolution spelling out automatic consequences for any state that was found in non-compliance with its safeguards agreement. Pierre Goldschmidt, for instance, proposes that the Security Council require non-compliant states to extend enhanced access rights to the IAEA and, if the IAEA Director General is not satisfied with the progress of an investigation, demand that all sensitive fuel-cycle activities be suspended.[48] However, regardless of the exact terms of such a resolution, any agreement on one would emphasise the futility of playing the intent card.

The US should apply a universal standard to friends as well as adversaries

Finally, and most importantly, the United States should be willing to lead by example and apply a universal standard to its friends as well as its adversaries. Washington has a pivotal role to play in preventing proliferation because the next proliferator is likely to be a US ally. Egypt, Saudi Arabia and Turkey – three of the states widely considered most likely to proliferate – are all friends of the United States. The same is true of many states that might seek nuclear weapons over the longer term, including Japan, South Korea, Argentina and Brazil (not to mention Taiwan). It is vital that these states believe that if they proliferate and are detected, an attempt to avoid any consequences by arguing benign intent will be unsuccessful. It is therefore imperative for the United States to make it clear that if a friend violates its non-proliferation obligations, Washington will permit a proportionate response, irrespective of the friend's claimed intent. For minor violations this would entail the Board of Governors referring the case to the Security Council, which might then choose to censure the state. In more serious cases, it would involve the Security Council implementing sanctions.

The United States scored an 'own goal' in its handling of South Korea's non-compliance. The country acknowledged in 2004 that it had conducted some undeclared enrichment experiments in 2000. This prompted an investigation by the IAEA that uncovered evidence of other clandestine nuclear activities, including reprocessing experiments. These activities constituted violations of South Korea's safeguards agreement.[49] In the run-up to the crucial Board of Governors meeting in November 2004, Seoul launched a major diplomatic offensive, arguing that the experiments had not been authorised by the government and that South Korea had no intention of building nuclear weapons.[50] This gave US officials, who were subject to desperate lobbying by the Korean government and who were already reluctant to refer an ally to the Security Council, the political space to argue that it was inappropriate to make a formal finding of non-compliance. In the end, the Board of Governors stopped short of finding South Korea in non-compliance and did not refer it to the Security Council.

The United States' failure to support South Korea's referral to the Security Council was a lost opportunity for non-proliferation. As its desperate lobbying demonstrates, Seoul would have considered referral a serious form of censure in itself and, given the nature of the non-compliance, no further consequence would have been necessary. Permitting a friend or ally to be referred to the Security Council has a cost, but this must be weighed against the benefits of deterring proliferation. By arguing against referral, the United States sent a signal to its friends that should one of them violate its non-proliferation obligations but claim (truthfully or not) that its intentions were benign, it would be spared any consequences. This undermined the deterrence value of safeguards and, ultimately, reduced the ability of the United States to prevent proliferation.

Notes

1 For a discussion of the procedure for finding a state in non-compliance and the cases of Iran, Egypt, Libya and South Korea see the article by Pierre Goldschmidt in this issue.

2 IAEA, 'Implementation of the NPT Safeguards Agreement in the Arab Republic of Egypt', GOV/2005/9, 14 February 2005, http://www.carnegieendowment.org/static/npp/Egypt_Feb_2005.pdf.

3 The IAEA's reports on Iran are available at http://www.iaea.org/NewsCenter/Focus/IaeaIran/index.shtml. For an assessment of the Iranian case see Mark Fitzpatrick, *The Iranian Nuclear Crisis: Avoiding Worst-Case Outcomes*, Adelphi Paper 398 (London: Routledge for the IISS, 2008).

4 IAEA, 'Implementation of the NPT Safeguards Agreement in the Republic of Korea', GOV/2004/84, 11 November 2004.

5 IAEA, 'Implementation of the NPT Safeguards Agreement in the Syrian Arab Republic', GOV/2008/60, 19 November 2008; David Albright and Paul Brannan, 'The Al Kibar Reactor: Extraordinary Camouflage, Troubling Implications', ISIS Report, 12 May 2008, http://www.isis-online.org/publications/syria/SyriaReactorReport_12May2008.pdf. The video by the US Intelligence Community that contained the first publicly released evidence of the reactor programme can be found at http://www.veoh.com/videos/v7050611ztAGtFDJ.

6 For instance, in casting the only vote against the adoption of UN Security Council Resolution 1696 (2006) – the first legally binding call for Iran to halt its enrichment programme – Qatar's ambassador pointed to uncertainty about Iranian intentions, saying, 'we would have seen no harm in waiting a few days so as to exhaust all possible ways and means in order to determine Iran's real intentions'. The importance of Iran's intentions was not disputed. Indeed, the British ambassador stated that 'we have given Iran many opportunities to show that it has no intention to develop nuclear weapons. Regrettably, Iran has not taken the steps required by the IAEA Board and the Security Council that would help build confidence.' United Nations Security Council, 5500th Meeting, S/PV.5500, 31 July 2006, pp. 3–4, available at http://www.un.org/Depts/dhl/resguide/scact2006.htm.

7 For the complete text of the treaty, see 'Treaty on the Non-Proliferation of Nuclear Weapons', INFCIRC/140, 22 April 1970.

8 Jozef Goldblat, *Arms Control: A New Guide to Negotiations and Agreements*

(London: Sage Publications, 2002), p. 102.

9 Quoted in full in Mohamed I. Shaker, *The Nuclear Non-Proliferation Treaty: Origin and Implementation 1959–1979*, Volume 1 (New York: Oceana Publications, 1980), p. 251 (emphasis added).

10 IAEA, 'Implementation of the NPT Safeguards Agreement and Relevant Provisions of Security Council Resolutions 1737 (2006) and 1747 (2007) in the Islamic Republic of Iran', GOV/2008/4, 22 February 2008.

11 United Nations Security Council, 5848th Meeting, S/PV.5848, 3 March 2008, pp. 7–8, available at http://www.un.org/Depts/dhl/resguide/scact2008.htm.

12 For instance, Joseph Cirincione (a non-proliferation expert) and Ray Takeyh (a Middle East expert) wrote that the IAEA 'is also satisfied that experiments with polonium-210 (that can be used as a trigger for an explosive nuclear chain reaction) were not part of a larger weapons project'. See Ray Takeyh and Joseph Cirincione, 'ElBaradei is Quietly Managing to Disarm Iran', *Financial Times*, 27 February 2008, http://www.ft.com/cms/s/0/06a1fa90-e4d7-11dc-a495-0000779fd2ac.html; for another example, which quotes an Iranian official promulgating the same interpretation, see Associated Press, 'Nuclear Watchdog Says Iran Rejects Evidence Linking it to Nuclear Weapons', *International Herald Tribune*, 22 February 2008, http://www.iht.com/articles/ap/2008/02/22/europe/EU-GEN-Nuclear-Iran.php.

13 IAEA, GOV/2008/4, para. 21.

14 *Ibid.*, para. 24 (emphasis added).

15 IAEA, 'Implementation of the NPT Safeguards Agreement in the Islamic Republic of Iran', GOV/2007/48, 30 August 2007, para. 9.

16 IAEA, GOV/2008/4, para. 11.

17 IAEA, 'Implementation of the NPT Safeguards Agreement and Relevant Provisions of Security Council Resolutions 1737 (2006) and 1747 (2007) in the Islamic Republic of Iran', GOV/2007/58, 15 November 2007, paras 8–23.

18 This interpretation of the IAEA's findings has been confirmed by a senior IAEA official. Private communication, Vienna, 2008. It would, however, probably be wrong to claim that the IAEA has *never* commented on Iranian intentions. For instance, given the Foster Criteria, the director general's notable but never-to-be-repeated 2003 comment that 'there is no evidence that the previously undeclared nuclear material and activities … were related to a nuclear weapons programme' would seem to be a comment on how he perceived Iran's intentions. IAEA, 'Implementation of the NPT Safeguards Agreement in the Islamic Republic of Iran', GOV/2003/75, 10 November 2003, para. 52. (See also note 36.) Similarly, in regard to Iran's centrifuge programme, the IAEA did state that it was 'not in a position, based on the information currently available to it, to draw conclusions about the original underlying nature of parts of the programme', which suggests that it might be in such a position in future. IAEA, GOV/2007/58, para. 7.

19 Jonathan Tirone and Matthias Wabl, 'UN Won't Refer South Korea Nuclear Activity to Security Council', Bloomberg, 26 November 2004, http://www.bloomberg.com/apps/news?pid=10000103&sid=a..54F3Aqem4&refer=us.

20 IAEA, GOV/2004/84, para. 41.

21 IAEA, 'The Structure and Content of Agreements Between the Agency and States Required in Connection with the Treaty on the Non-Proliferation of Nuclear Weapons', INFCIRC/153 (Corrected), June 1972, para. 28.

22 Mohamed ElBaradei, 'Nuclear Non-proliferation and Arms Control: Are We Making Progress?', Remarks at the 2005 Carnegie International Nonproliferation Conference, Washington DC, 7 November 2005, www.iaea.org/NewsCenter/Statements/2005/ebsp2005n017.html; for a detailed discussion of the IAEA's right to inspect suspected weaponisation activities see James Acton with Carter Newman, *IAEA Verification of Military Research and Development*, Verification Matters 5 (London: Verification Research, Training and Information Centre (VERTIC), 2006), pp. 11–18, http://www.vertic.org/publications/VM5%20(2).pdf.

23 Russia, in particular, has stressed the need for 'proof' or 'certainty'. See, for instance, 'Block on Iran Sanctions: Russia Demands Proof of Hidden Program', *International Herald Tribune*, 22 April 2006, http://www.iht.com/articles/2006/04/21/news/russia.php; 'Ahmadinejad Declares Victory in Nuclear Dispute', *Times Online*, 5 December 2007, http://www.timesonline.co.uk/tol/news/world/us_and_americas/article3004070.ece. For a discussion of the effect that an insistence on proof has on enforcing non-proliferation agreements see George Perkovich and James M. Acton, *Abolishing Nuclear Weapons*, Adelphi Paper 396 (Abingdon: Routledge for the IISS, 2008), p. 75.

24 Then IAEA Deputy Director General for Safeguards, Pierre Goldschmidt, wrote about Libya that if it 'had not admitted that it had a nuclear weapons programme, the Agency would not have been in a position to prove that the uranium conversion and enrichment activities undertaken covertly by Libya over more than 20 years were for such a programme'. Pierre Goldschmidt, 'Reinforcing Nuclear Safeguards and the Role of the IAEA', High Level Seminar of Weapons of Mass Destruction, Brussels, 17 March 2005, http://www.carnegieendowment.org/publications/index.cfm?fa=view&id=22328.

25 James Acton with Carter Newman, *IAEA Verification of Military Research and Development*, pp. 26–33.

26 Using either the special inspection provision of the Comprehensive Safeguards Agreement or Article 5.c complementary access (if an Additional Protocol is in force).

27 UN Security Council Resolution 707 (1991), para. 3(b); UN Security Council Resolution 1441 (2002), para. 5.

28 IAEA, 'Implementation of the NPT Safeguards Agreement in the Islamic Republic of Iran', GOV/2005/87, 18 November 2005, para. 6.

29 IAEA, 'Implementation of the NPT Safeguards Agreement and Relevant Provisions of Security Council

Resolutions 1737 (2006), 1747 (2007) and 1803 (2008) in the Islamic Republic of Iran', GOV/2008/38, 15 September 2008, para. 17(d).

30 IAEA, 'Implementation of the NPT Safeguards Agreement and relevant provisions of Security Council resolutions 1737 (2006), 1747 (2007), 1803 (2008) and 1835 (2008) in the Islamic Republic of Iran', GOV/2008/59, 19 November 2008, paras. 15–17.

31 Article XII.A.6 of the IAEA's statute does envision inspectors having the right to access 'any person who by reason of his occupation deals with materials, equipment, or facilities which are required by this Statute to be safeguarded'. However, the legal principle of *lex specialis derogat legi generali* (a more specific law supersedes a more general one) means that this right is not enforceable as it is not contained in the subsequent and much more detailed Comprehensive Safeguards Agreement.

32 IAEA, GOV/2008/59, para. 16; Pierre Goldschmidt, 'IAEA Safeguards: Dealing Preventively with Non-Compliance', 12 July 2008, pp. 8–9, http://www.carnegieendowment.org/files/Goldschmidt_Dealing_Preventively_7-12-08.pdf.

33 Goldschmidt, 'IAEA Safeguards: Dealing Preventively with Non-Compliance', pp. 8–9.

34 UN Security Council Resolution 1441 (2002), para. 5. Such a provision did not appear in Resolutions 687 (1991) or 707 (1991) that provided the initial mandate for IAEA investigations in Iraq.

35 Annette Berriman, Russell Leslie and John Carlson, 'Assessing Motivation as a Means of Determining the Risk of Proliferation', paper presented to the 45th Annual Meeting of the Institute of Nuclear Materials Management, Orlando, FL, 18–22 July 2004, http://www.asno.dfat.gov.au/publications/inmm2004_motivation.pdf.

36 A good example of this is the director general's claim from 2003 that there was 'no evidence' of a nuclear-weapons programme in Iran (see note 18). The fact that in the following sentence he added 'given Iran's past pattern of concealment, it will take some time before the Agency is able to conclude that Iran's nuclear programme is exclusively for peaceful purposes' would strongly suggest that there was indeed evidence (if not proof). IAEA, GOV/2003/75, para. 52. Moreover, according to recent press reports, the director general attempted to make the IAEA's February 2008 report more positive than his technical staff believed was warranted. 'UN Nuclear Watchdog Could Delay Iran Nuclear Report', *Global Security Newswire*, 12 February 2008, http://www.nti.org/d_newswire/issues/2008_2_12.html.

37 I am very grateful to Jeffrey Lewis of the New America Foundation for pointing this out to me. See also Bruno Tertrais, 'Has Iran Decided to Build the Bomb? Lessons from the French Experience', Proliferation Analysis, Carnegie Endowment for International Peace, 30 January 2007, http://www.carnegieendowment.org/publications/index.cfm?fa=view&id=18993.

38 Paul M. Cole, *Atomic Bombast: Nuclear Weapon Decisionmaking in*

Sweden 1945–1972, Occasional Paper 26 (Washington DC: The Henry L. Stimson Center, 1996), pp. 19–26, http://www.stimson.org/wmd/pdf/cole.pdf.

39 *Ibid*, p. 21.

40 Eric Arnett, 'Norms and Nuclear Proliferation: Sweden's Lessons for Assessing Iran', *Nonproliferation Review*, vol. 5, no. 2, Winter 1998, p. 35.

41 Pursuant to Article XII.C of the Statute of the IAEA, the Board of Governors also has a few modest enforcement powers such as the suspension of technical cooperation, which it recently exercised with respect to Iran and was called upon to use in the case of Syria in November 2008, although it declined to do so. See Mark Heinrich, 'IAEA Chief, West Clash over Nuclear Aid for Syria', Reuters, 24 November 2008, http://www.reuters.com/article/worldNews/idUSTRE4AN5BB20081124?rpc=401&.

42 This essay does not criticise such practices. Its focus is on the work of the multinational-bodies strand.

43 See note 23.

44 United Nations Security Council, 5848th Meeting, S/PV.5848, 3 March 2008, p. 21, available at http://www.un.org/Depts/dhl/resguide/scact2008.htm.

45 This objection was raised several times during discussions about the ideas raised in this essay prior to publication.

46 This framework is not authoritative or definitive, but is a first attempt to demonstrate that it is possible to distinguish between instances of non-compliance without focusing on intent.

47 Very little information about this event is in the public domain. See, for instance, David Fischer, *History of the International Atomic Energy Agency: The First Forty Years* (Vienna: IAEA, 1997), p. 321, fn. 152, http://www-pub.iaea.org/MTCD/publications/PDF/Pub1032_web.pdf.

48 Goldschmidt, 'IAEA Safeguards: Dealing Preventively with Non-Compliance'.

49 The IAEA's findings are published in IAEA, GOV/2004/84.

50 Dafna Linzer and Anthony Faiola, 'US Won't Report South Korea to UN for Nuclear Tests', *Washington Post*, 25 November 2004, p. A19. Indeed, as part of this diplomatic offensive, then Korean Foreign Minister (now UN Secretary-General) Ban Ki Moon wrote an op-ed outlining his country's defence in the *Washington Post*. Ban Ki Moon, 'South Korea's Nuclear Safeguards', *Washington Post*, 8 October 2004, p. A35, http://www.washingtonpost.com/wp-dyn/articles/A16508-2004Oct7.html.

Exposing Nuclear Non-compliance

Pierre Goldschmidt

The nuclear non-proliferation regime, so vital to maintaining international peace and security, is under increasing threat, particularly from countries that deliberately violate their non-proliferation obligations. Experience with North Korea and Iran has demonstrated that non-compliance must be addressed promptly and effectively. Iran has sought to exploit inconsistencies in how the International Atomic Energy Agency (IAEA) reports violations, including its own case and that of Libya, as well as the less worrying but still significant cases of South Korea and Egypt. Clarifying the technical and statutory basis by which the IAEA exposes non-compliance is one immediate way the non-proliferation regime can be strengthened.

It is hard to believe that, more than 35 years after the adoption of the model Comprehensive Safeguards Agreement, the meaning of 'non-compliance' is still uncertain and subject to debate.[1]

Over the last few years, several questions have been repeatedly raised: What is non-compliance? How does one distinguish non-compliance from 'minor reporting oversights'? What happens if the 'mistake' is a one-time incident? Who decides that a state is in non-compliance? Were South Korea and Egypt found to be in non-compliance in 2004 and 2005, and if so, why were they not reported to the UN Security Council? It is time for the IAEA Board of Governors to address these questions, set the record straight, and

Pierre Goldschmidt is Non-resident Senior Associate at the Carnegie Endowment for International Peace, and former Deputy Director General of the International Atomic Energy Agency, Head of the Department of Safeguards.

Survival | vol. 51 no. 1 | February–March 2009 | pp. 143–164 DOI 10.1080/00396330902749764

assure that any future non-compliance is recognised and responded to consistently and effectively.

There is a broad consensus that states which violate or withdraw from the Non-Proliferation Treaty should not be allowed to benefit from the nuclear-energy assistance they received (under Article IV) while and because they were a party to the treaty, or to profit from their violation with impunity. It has therefore been recommended[2] that the Security Council should adopt a generic (non state-specific) and legally binding resolution, under Chapter VII of the UN Charter, stating that if a state notifies its withdrawal from the treaty *after* being found by the agency to be in non-compliance with its safeguards undertakings, such withdrawal notice constitutes a threat to international peace and security and the state in question would have to surrender all materials and equipment it received under its safeguards agreement(s).

I have also suggested that the Security Council should adopt another generic resolution providing that, if a state is found to be in non-compliance and does not fully and proactively cooperate with the IAEA in resolving the issue and taking the necessary corrective actions, the non-compliant state would be obliged to temporarily suspend all sensitive nuclear-fuel-cycle activities (in particular those related to uranium enrichment, spent fuel reprocessing and the separation of plutonium).[3]

These generic resolutions would not affect states in good standing with their non-proliferation undertakings. It is therefore essential to be clear about what constitutes non-compliance and how the IAEA is to identify and deal with such cases.

Reporting non-compliance

According to Article XII.C of the IAEA Statute, reporting a state to the Security Council for non-compliance with its safeguards undertakings can be seen as a process comprising the following steps, the last three of which can be taken in sequence or simultaneously:

- Step 1: Agency inspectors report any non-compliance to the director general, via the deputy director general for safeguards.

- Step 2: The director general transmits the report to the Board of Governors.
- Step 3: The board makes a formal finding of non-compliance.
- Step 4: The board calls upon the state in question 'to remedy forthwith any non-compliance which it finds to have occurred'.
- Step 5: The board reports the non-compliance to all members and to the Security Council and General Assembly of the United Nations.

Since 2003, the IAEA Secretariat has reported specific cases of non-compliance with safeguards agreements by Iran, Libya, South Korea and Egypt to the board (Step 2). The actions taken by the board in each case were inconsistent and, if they go uncorrected, will create unfortunate precedents.

The case of Iran is particularly relevant. Iran claims that the decision taken by the Board of Governors in February 2006 to report Iran's nuclear issue to the UN Security Council 'had no legal basis and contradicted the IAEA Statute and its practice'.[4] Iran bases its case on a claim that 'the inspectors have not reached or concluded any "non- compliance"'.[5] These statements are incorrect,[6] and need to be rebutted officially, preferably by the IAEA Legal Office. Because no such rebuttal has yet been made, confusion and misunderstandings have arisen due to the language used in reporting the findings made by the IAEA Secretariat. The confusion is compounded by the board's responses to safeguards breaches involving South Korea and Egypt which were effectively cases of non-compliance.

These statements are incorrect and need official rebuttal

Whether or not the word 'non-compliance' is used in the report transmitted to the board in Step 2 is irrelevant, as demonstrated in the case of Libya, which admitted to working on an undeclared nuclear-weapons programme for many years. This was an indisputable case of non-compliance with Libya's Non-Proliferation Treaty and safeguards undertakings. However, in the director general's report to the board in February 2004, the word 'non-compliance' was not used; rather, it was stated that 'Libya was in breach of its obligation to comply with the provisions of the Safeguards Agreement', which is synonymous.[7] Certainly to be 'in breach of one's obligations to comply' and to be in 'non-compliance' is a distinction without a difference.

The same language was used in the director general's November 2003 report on Iran, in which he stated that 'in the past, Iran has concealed many aspects of its nuclear activities, with resultant breaches of its obligation to comply with the provisions of the Safeguards Agreement' which 'has given rise to serious concerns'.[8]

The director general's special report on South Korea in November 2004 found that on a number of occasions the country failed to report to the agency experiments and activities involving nuclear material:

a. Failure to report nuclear material used in evaporation, spectroscopy and enrichment experiments ([atomic vapour laser isotope separation] and chemical exchange) and the associated products;
b. Failure to report the production, storage and use of [natural uranium] metal and associated process loss of nuclear material, and the production and transfer of waste resulting therefrom;
c. Failure to report the dissolution of an irradiated mini-assembly and the resulting uranium–plutonium solution, including the production and transfer of waste; and
d. Failure to report initial design information for the enrichment facilities and updated design information for the facilities involved in the plutonium separation experiment and the conversion to [natural uranium] and [depleted uranium] metal.[9]

In the case of Egypt, the director general's February 2005 report to the board summarised the country's failure to report a number of nuclear material and activities:

a. Failure to report on its initial inventory imported UF_4, imported and domestically produced uranium metal, imported thorium compounds, small quantities of domestically produced UO_2, UO_3 and UF_4, and a number of unirradiated low enriched and natural uranium fuel rods;

b. Failure to report the uranyl nitrate and scrap UO_2 pellets, and their use for acceptance testing of the Hydrometallurgy Pilot Plant;
c. Failure to report the irradiation of small amounts of natural uranium and thorium and their subsequent dissolution in the Nuclear Chemistry Building laboratories, including the production and transfer of waste;
d. Failure to provide initial design information for the Hydrometallurgy Pilot Plant and the Radioisotope Production Facility, and modified design information for the two reactors.[10]

Instead of referring to 'breaches of its obligation to comply' as the reports for Iran and Libya did, these reports use the following language: 'The ROK conducted experiments and activities involving uranium conversion, uranium enrichment and plutonium separation, which it *failed to report* to the Agency *in accordance with its obligation* under its Safeguards Agreement'[11] and these failures were 'a matter of serious concern'.[12] In the case of Egypt, 'the Agency has identified a number of *failures by Egypt to report* to the Agency *in accordance with its obligations* under its Safeguards Agreement';[13] and 'the repeated failures by Egypt to report nuclear material and facilities to the Agency in a timely manner are *a matter of concern*'.[14]

The question has therefore been raised about exactly what constitutes non-compliance under Article XII.C of the IAEA Statute. Clearly a failure to declare nuclear material and activities that should be subject to IAEA safeguards cannot be considered a small, technical reporting mistake. If such failures have taken place over an extended period of time or were deliberately concealed they must be categorised as non-compliance. If these activities had a plausible military purpose or involved military organisations then they are an even greater matter of concern. The same is true if the non-compliant state does not fully and actively cooperate with the IAEA to remedy the situation.

Without attempting to be exhaustive, the following are clear cases of non-compliance under the Comprehensive Safeguards Agreement (especially if they are combined):

- 'Diversion of nuclear material from declared nuclear activities, or the failure to declare nuclear material required to be placed under safeguards'[15] (for example not declaring nuclear material in the initial inventory or imported thereafter).

- Undeclared production of nuclear material (for example through irradiation).
- Undeclared use of exempted nuclear material in a nuclear facility.
- 'Obstruction of the activities of IAEA inspectors, interference with the operation of safeguards equipment, or prevention of the IAEA from carrying out its verification activities'.[16]
- Starting the construction[17] or modifying the design of a nuclear facility without informing the agency.

When an Additional Protocol is in force, many more actions would constitute non-compliance, such as 'the failure to declare nuclear material, nuclear activities or nuclear related activities required to be declared under Article 2' or 'the undeclared manufacture of items referred to in Annex I' of the Additional Protocol.[18]

The non-proliferation regime would be significantly strengthened if the board were to confirm that any state-specific report on safeguards implementation (beyond the annual Safeguards Implementation Report[19]), transmitted by the director general to the board, is to be considered, unless explicitly stated otherwise, as a report of non-compliance (or a progress report on verification activities following a case of non-compliance).

Aside from clear cases of non-compliance, the extent to which an accumulation of safeguards-implementation problems – such as delayed provision of access to nuclear material or facilities, material unaccounted for, open questions and unresolved inconsistencies, inconclusive containment and surveillance results, and lack of cooperation with the agency – justifies a state-specific report to the board should be left to the judgement, and is the responsibility of, the IAEA Department of Safeguards and the director general. However, the board could insist that all such problems be reported in a sufficiently transparent way in the Safeguards Implementation Report. This could include the names of the states where such problems have occurred without being resolved in a timely manner. It would allow the board, if it deems necessary, to request more information from the secretariat on any specific case of potential proliferation concern.

Where the director general has reported a case of technical and legal non-compliance, the board is not obliged to make a formal finding of non-compliance if it judges the circumstances to not warrant it. If the board does decide to confirm its finding (Step 3) that the state under review is in non-compliance with its safeguards undertakings, it will normally do so by adopting a state-specific resolution. Accordingly, in a March 2004 resolution the board declared Libya to be in non-compliance and requested the director general to report the matter to the Security Council (Step 5), and in a September 2005 resolution declared Iran to be in non-compliance.[20] Both resolutions referred to Article XII.C of the IAEA Statute, while the resolution concerning Iran also referred to Article III.B.4.[21] In the cases of South Korea and Egypt the board did not adopt resolutions nor formally declare either state to be in non-compliance. In each case, it endorsed a 'Chairman's Conclusion' summarising the board's discussion of the issue and encouraging the states to continue their active cooperation with the agency.

The board has total freedom in formulating a resolution

In practice, reporting a state to the Security Council is achieved through a board resolution requesting the director general, on behalf of the board, to transmit the resolution to all members, the UN Security Council and the UN General Assembly. However, the board has total freedom in formulating the resolution as it deems appropriate. With regard to the timing of Step 5, the board is obliged to report a finding of non-compliance to the Security Council, but the Statute does not specify a time limit for the board to do so. For instance, when the board adopted a resolution in September 2005 finding Iran to be in 'non-compliance in the context of Article XII.C of the Agency's Statute' it also stated that 'the Board will address the timing and content of the report required under Article XII.C and the notification required under Article III.B.4'.[22]

The main reason the board did not report Iran to the Security Council in November 2003 (when the director general submitted a comprehensive and damning report) was that Iran had informed the agency on 10 November

2003 (the day the director general's report was issued) that it had agreed to sign the Additional Protocol (which it did on 18 December 2003) and to suspend all enrichment-related and reprocessing activities, not to produce feed material for enrichment processes and not to import enrichment-related items.[23] On 21 October 2003, the foreign ministers of France, Germany and the United Kingdom (the E3) had visited Iran to discuss the nuclear issue. At the end of that visit Iran had agreed to cooperate fully with the IAEA to settle all outstanding issues and to correct any failures to comply with its safeguards agreement. To that end, Iran announced its willingness to sign an Additional Protocol and to commence the ratification process. Iran also agreed to 'voluntarily suspend all uranium enrichment and reprocessing activities as defined by the IAEA'. On their part, the three European foreign ministers informed Iran that 'in their view, full implementation of Iran's decisions, confirmed by the IAEA's Director General, should enable the immediate situation to be resolved by the IAEA Board'.[24] This sentence was understood as meaning that if Iran complied with its commitments the E3 would not seek Iran's referral to the Security Council at the November meeting of the board.

In the case of Libya, the resolution adopted in March 2004 by the board requested the director general 'to report the matter to the Security Council *for information purposes only,* while commending [Libya] for the actions it has taken to date, and has agreed to take, to remedy the non-compliance' (emphasis added). The Security Council held a meeting on 22 April 2004 to consider the matter, and its president made a statement on behalf of the council welcoming Libya's active cooperation with the IAEA and its decision to abandon its weapons of mass destruction programmes.[25]

This was a clear demonstration that reporting non-compliance to the board, and subsequently to the Security Council, does not necessarily entail sanctions of any kind. Quite the contrary:

- In April 2004 the United States terminated the applicability of the Iran–Libya Sanction Act to Libya.
- In July 2004 Libya re-established its diplomatic presence in Washington.

- In June 2005, France became the first nuclear-weapon state to consider civilian nuclear cooperation with Libya. The scope of such an agreement mainly involved the reactivation of Libya's Tajura nuclear research reactor. In July 2007, a Memorandum of Understanding was signed envisioning collaboration on a desalination plant with one or more French nuclear reactors as well as support in exploration for and production of uranium. An agreement was signed in a meeting held in December 2007.[26]
- In June 2006 the United States formally rescinded Libya's designation as a state sponsor of terrorism.
- In January 2008 Libya took its seat as a member of the UN Security Council.

Finally, the IAEA September 2008 report on Libya states that, since the disclosure in December 2003 of its undeclared nuclear activities, 'Libya has provided the Agency unrestricted and prompt access, *beyond that required under its Safeguards Agreement and Additional Protocol,* to those locations, information and individuals deemed necessary by the Agency to fulfil its verification requirement'. The report then concludes that 'the Agency, in accordance with its procedures and practices, will *continue* to implement safeguards in Libya *as a routine matter*'.[27]

Some member states may object to the proposal that any state-specific report by the director general to the board is (unless explicitly stated otherwise) automatically deemed to be a report on non-compliance, because the IAEA Secretariat might then be less likely to report in that way any serious breach to the board, but would opt to report these breaches in a less transparent and detailed manner in the annual Safeguards Implementation Report. Yet the risk that the secretariat would downplay material failures by any state to comply with its safeguards undertakings seems very low for several reasons. First, although in some cases there will inevitably be a question of judgement about whether or not to report failures and breaches in a state-specific report to the board, the secretariat cannot, and will not, take the risk of covering up a situation that might turn out, later on, to have been an indication of undeclared nuclear activities, thereby undermining the credibility of IAEA safeguards. Secondly, as demonstrated in the Libyan case, the secretariat and the board can qualify the non-compliance as they deem appropriate, taking into account corrective actions taken and cooperation provided by the state under review. In particular, the board can decide, as with Libya, to report the matter to the Security Council 'for information purposes only' with no negative consequence for the state in question.

Two unfortunate precedents

On 26 November 2004, the board decided not to adopt a resolution on South Korea and, therefore, not to report the case to the Security Council, setting an unfortunate precedent motivated at least in part by political considerations. Nevertheless, the chairman of the IAEA Board concluded that 'the Board shared the Director General's view that given the nature of the nuclear activities described in his report, the failure of the Republic of Korea to report these activities in accordance with its safeguards agreements is of *serious concern*'.[28] Since the board is obliged to report any case of non-compliance to the Security Council, not doing so in the case of South Korea could be interpreted as meaning that the board did not consider the breaches to constitute non-compliance with Comprehensive Safeguards Agreements.

Among the reasons for not reporting South Korea to the Security Council, one can highlight the fact that Seoul took the initiative of informing the secretariat that it had discovered, in June 2004, in connection with the submission of its initial declaration pursuant to the Additional Protocol, that laboratory-scale experiments had been carried out by scientists at the Korea Atomic Energy Research Institute (KAERI) in Daejon. South Korean officials subsequently launched a major diplomatic offensive insisting that the government was not aware of, and did not authorise, these experiments. According to the chairman, 'the Board welcomed the corrective actions taken by the Republic of Korea, and the active cooperation it has provided to the Agency' in providing timely information and access to personnel and locations. Moreover, 'the Board noted that the quantities of nuclear material involved have not been significant, and that to date there is no indication that undeclared experiments have continued'.[29] Political considerations also played a dominant role in the board's decision. At the time, the much more severe violations committed by Iran had not yet been formally declared by the board to constitute non-compliance,[30] and reporting South Korea would have been politically embarrassing since Seoul was a member of the Six-Party Talks underway to resolve the crisis created by North Korea's (the Democratic Peoples' Republic of Korea, DPRK) withdrawal from the Non-Proliferation Treaty in January 2003.

Thanks to Seoul's full cooperation in implementing the Additional Protocol, the secretariat was able to report in the Safeguards Implementation Report for 2007 that it had found no indication of the diversion of declared nuclear material from peaceful activities and no indication of undeclared nuclear material or activities. The secretariat has concluded that *all* nuclear material in South Korea remained in peaceful activities.[31]

It is clear, nonetheless, that South Korea was in non-compliance with its safeguards agreement since, in addition to the failures already mentioned, the November 2004 report to the board clearly shows that a number of sensitive activities involving undeclared nuclear material had taken place over an extended period of time and that South Korea initially took some actions which could be interpreted as attempts to conceal past failures.[32]

In the case of Egypt, in his February 2005 report to the board the director general stated that 'to date, the Agency has identified a number of failures by Egypt to report to the Agency in accordance with its obligations under its Safeguards Agreement' and concluded that

> the research and development activities referred to in this report were the subject of [Egyptian Atomic Energy Authority] and other scientific publications. Notwithstanding, and irrespective of the current status of the previously undeclared activities and the small amounts of nuclear material involved, the repeated failures by Egypt to report nuclear material and facilities to the Agency in a timely manner *are a matter of concern*.[33]

Nevertheless, the board decided neither to adopt a resolution nor to report the matter to the Security Council, and did not request that the report be made public, as had been done in the case of South Korea.

In its decision not to report Egypt to the Security Council, it is likely that the board took into account the South Korean precedent, its wish not to put Egypt in the same category as Iran, Egypt's cooperation with the agency, and the corrective actions it had already taken. In addition, it seems that the board considered that the nature of the undeclared activities described in the director general's report was not a matter of proliferation concern, taking into account that some of these activities had been the subject of open-source publications, that some of the activities took place 15–40 years earlier and that the amount of undeclared nuclear material involved was small.

None of these possible justifications can be considered satisfactory. Indeed, many of the undeclared activities referred to in the report are of a sensitive nature, since they include the irradiation of uranium and thorium targets dissolved in three laboratories that had not been declared to the agency. Moreover, Egypt failed to declare imported un-irradiated fuel rods containing uranium enriched to 10% U-235, some of which had been used in undeclared fuel-dissolution experiments, said to have been carried out prior to entry into force of Egypt's safeguards agreement. The fact that some (but not all) of the undeclared activities were the subject of open-source publications cannot be a valid justification for not reporting them to the agency. Otherwise, the mere publication somewhere of something about nuclear experiments by a state engaged in clandestine nuclear activities would permit that state to claim a 'minor reporting oversight' whenever such activities were discovered.

Although some of the undeclared experiments took place decades ago, the uranium and thorium irradiation and dissolution experiments were conducted between 1990 and 2003 (about when the agency started its investigation). It would have been as accurate to say that Egypt has used undeclared nuclear material in undeclared activities over a period of more than 20 years as to say that the country's undeclared activities were minor because some took place so long ago.

That Egypt failed to declare 67kg of imported uranium tetrafluoride (UF_4), 3kg of uranium metal, 9.5kg of imported thorium compounds, un-irradiated fuel rods containing 10% enriched U-235, and irradiation and dissolution experiments should not be considered a 'minor breach' of its safeguards agreement. These activities are highly significant, and treating them as minor would set a detrimental new standard. Iran used 'only' 1.6kg uranium hexafluoride (UF_6) for testing centrifuges at Kalaye Electric, and used about 400kg of UF_4, 400kg of uranium dioxide (UO_2) and 30kg of uranium metal in a number of undeclared experiments, which were not considered trivial by the board although the quantities involved could also be considered 'small'.

The IAEA Annual Report for 2004 mentions with regard to Egypt that 'in 2004, the Agency identified several open-source documents, which indicated the possibility of hitherto unreported nuclear material, activities and facilities in that State', and that 'at the end of 2004, the Agency was still in the process of verifying the correctness and completeness of Egypt's declarations'.[34] Although a full report on Egypt was submitted to the board in February 2005, there is no mention of Egypt in the IAEA Annual Report for 2005, or in any later one, and then not even in the Safeguards Implementation Report for 2007, giving the false impression that this issue had been resolved.[35]

The cases of Egypt and Iran are very different, in particular because for Egypt (and South Korea) there was no indication of military involvement and no grounds for concluding that there had been a systematic effort to evade safeguards obligations or that a policy of concealment had existed. But there is a danger of setting bad precedents based on arbitrary criteria or judgements informed by political considerations. For its part, according to the agency, 'Egypt has explained that its past failure to report was attributable to a lack of clarity about its obligations under its safeguards agreement, particularly as regards small quantities of nuclear material used in research and development activities'.[36] After the extensive reporting on Iran and Libya this appears to be a weak excuse, and such undeclared nuclear material and activities constitute clear cases of non-compliance.

There is a danger of setting bad precedents

It would also be a major mistake for member states to consider that because all the undeclared activities uncovered by the IAEA in Egypt are permissible under the Non-Proliferation Treaty if declared and placed under safeguards, they should not be a matter of concern. Such a conclusion would negate the whole purpose of comprehensive safeguards agreements. It is true that, taken individually, the failures committed by Egypt would most likely not justify being reported to the board outside the annual Safeguards Implementation Report. However, because of their number and the extended period over which these failures have taken place, a special report to the board was justified.

The dilemma for the board is to decide whether or not these failures constitute non-compliance under Article XII.C of the Statute. From a technical and legal point of view there is no doubt that this is indeed the case. If the board, however, considers that more information is necessary to make such a finding, it should have long since requested the director general to provide an update of his February 2005 report. It is now almost four years since it was reported that 'the Agency's verification of the correctness and completeness of Egypt's declarations is ongoing, pending further results of environmental and destructive sampling analyses and the Agency's analysis of the additional information provided by Egypt'.[37]

Can the damage be repaired?

To repair the damage done by not reporting non-compliance by South Korea and Egypt to the Security Council, the board should decide that, from now on, any state-specific report by the IAEA Secretariat to the board on safeguards-implementation issues, unless the report explicitly states otherwise, shall henceforth be deemed to be a report of non-compliance (or a progress report on verification activities following a case of non-compliance), whether or not the words 'non-compliance' are used in the report. For example, the most recent report on the implementation of safeguards in Syria[38] could be considered a report of non-compliance as foreseen under Article 19 of the safeguards agreement,[39] since Syria denied the agency the requested access to locations, documentation and information, preventing the agency from carrying out without delay its verification activities, including the determination of the origin of the anthropogenic natural-uranium particles found at the Dair Alzour site. Moreover, Syria undertook possible concealment activities such as large-scale clearing, levelling and landscaping operations and the removal of large containers shortly after the agency's request for access.

If the secretariat considered the information available and Syria's refusal of access to a number of suspected locations insufficient to constitute non-compliance, the final section of the report to the board relating to the 'Current Assessment' could have included a paragraph stating that the present report was not a report of non-compliance by Syria with its safeguards obligations but on a possible case of non-compliance that must be verified without further delay by the agency.

The agency has the legal authority, under INFCIRC/153-type safeguards agreements, 'to conduct special inspections insofar as these relate to the verification of the existence or non-existence of undeclared activities'.[40] The board should have adopted a resolution in November 2008 requiring Syria to provide immediate access to all relevant information, documentation and locations and specifying that if it did not comply within 30 days, the board, acting under Article 18 of Syria's safeguards agreement, would request the director general to make a special inspection as provided under Article 73 of Syria's safeguards agreement.[41] As the director general stated in January 1992, the agency has 'the authority, under the Statute and under comprehensive safeguards agreements concluded with it, to request special inspections at undeclared sites'.[42] In any case, 'in considering a question of non-compliance, the Board might, on its own initiative, conclude that the circumstances warrant a special inspection'.[43]

Independently of the decision of principle to consider state-specific safeguards reports as reports of non-compliance, the board should bring the South Korean and Egyptian cases into conformity with this new standard. The board should therefore adopt a resolution acknowledging that the failure by Seoul to declare a number of experiments and activities involving nuclear material as reported to the board in November 2004 constitutes non-compliance with its safeguards agreement (in the context of Article XII.C of the Statute), commending South Korea for its cooperation with the agency in providing access to information, documents, persons and locations, welcoming the fact that all nuclear material in South Korea remained in peaceful activities, and requesting the director general to report this resolution and all reports and chairman's conclusions relating to South Korea to the Security Council for information purposes only.[44]

Similarly, the board should adopt a resolution acknowledging that Egypt's failure to report a number of nuclear materials and activities, as reported by the board in February 2005, constitutes non-compliance with its safeguards agreement (in the context of Article XII.C of the Statute), commending Egypt for its cooperation with the agency, underlining the necessity for Egypt to sign and ratify the Additional Protocol promptly and, in the meantime, to fully implement its provisions on a voluntary basis at least until such time as the IAEA Secretariat has concluded that Egypt's declarations are correct and complete.[45] The board resolution should also request the director general to report this resolution and all reports and chairman's conclusions relating to Egypt to the Security Council for information purposes only, and to provide an updated report on its verification findings in Egypt.

Finally, the board should request the secretariat to report more explicitly on borderline cases in the annual Safeguards Implementation Report, with the names of the states concerned and a description of the difficulties faced in implementing its verification activities or any finding that may potentially raise proliferation concerns. Experience has shown that disclosure in this report of the names of states that are not fully cooperating with the secretariat has often had a positive effect.[46] This practice should be continued and expanded. It would allow the board, if it deems necessary, to request more information from the secretariat on any specific case of possible proliferation concern.

* * *

In 2005 the IAEA was awarded the Nobel Peace Prize, to a large extent in recognition of its objectivity and impartiality when dealing with nuclear non-proliferation issues. In light of the irreplaceable role of the IAEA in this domain it is crucial that its reputation and the credibility of the safeguards regime be maintained. It is therefore necessary for the agency to formally acknowledge that in the past some of its decisions have created potentially damaging precedents that need to be corrected to avoid any impression that the implementation of the IAEA Statute is selective. Member states have repeatedly insisted that 'measures to strengthen the effectiveness and improve the efficiency of the safeguards system with a view to detecting undeclared nuclear material and activities must be implemented rapidly by all concerned States'.[47] The Board of Governors will go a long way towards achieving this goal if it can demonstrate through its actions that it will not shirk its responsibilities when it comes to non-compliance with safeguards agreements.

Acknowledgements

The views expressed herein are those of the author only.

Notes

1 For the Comprehensive Safeguards Agreement see IAEA, 'The Structure and Content of Agreements Between the Agency and States Required in Connection with the Treaty on the Non-Proliferation of Nuclear Weapons', INFCIRC/153 (Corrected), June 1972. For the Additional Protocol, see 'Model Protocol Additional to the Agreement(s) between State(s) and the International Atomic Energy Agency for the Application of Safeguards', INFCIRC/540 (Corrected), http://www.iaea.org/Publications/Documents/Infcircs/1997/infcirc540c.pdf.

2 See, for example, George Perkovich, Jessica T. Matthews, Joseph Cirincione, Rose Gottemoeller and Jon B. Wolfsthal, *Universal Compliance: A Strategy for Nuclear Security* (Washington DC: Carnegie Endowment for International Peace, 2005), http://www.carnegieendowment.org/files/UC2.FINAL3.pdf, and Pierre Goldschmidt, 'Priority Steps to Strengthen the Nonproliferation Regime', Carnegie Endowment, Policy Outlook no. 33, February 2007, http://www.carnegieendowment.org/files/goldschmidt_priority_steps_final.pdf.

3 See, for example, Pierre Goldschmidt, 'The Increasing Risk of Nuclear Proliferation: Addressing the Challenge', Belgian Nuclear Society Symposium, Brussels, 26 November 2003, http://www.iaea.or.at/PrinterFriendly/NewsCenter/Statements/DDGs/2003/goldschmidt26112003.html, and Pierre Goldschmidt, 'IAEA Safeguards: Dealing Preventively with Non-compliance', Carnegie Endowment for International Peace, 12 July 2008, http://www.carnegieendowment.org/files/Goldschmidt_Dealing_Preventively_7-12-08.pdf.

4 See, for example, the statement by Reza Aghazadeh, Iran's vice president and president of the Atomic Energy Organisation of Iran, at the IAEA General Conference, 18–22 September 2006.

5 Letter from H.E. Ambassador Dr Soltanieh to the Director General of the IAEA, GOV/INF/2007/8.

6 Referring Iran to the Security Council was consistent with Articles XII.C and III.B.4 of the IAEA Statute and with its practice as demonstrated by the referral of Libya to the Security Council in 2004 on the basis of a report that did not use the word 'non-compliance'. Iran was clearly in non-compliance through failure, over many years, to declare all nuclear material and activities that should have been subject to safeguards, in circumstances where a military purpose was plausible; see Pierre Goldschmidt, 'Rule of Law, Politics and Nuclear Non-proliferation', presentation to the Ecole Internationale de Droit Nucléaire, Université de Montpellier, 7 September 2007, http://www.carnegieendowment.org/publications/index.cfm?fa=view&id=19564&prog=zgp&proj=znpp.

7 IAEA, 'Implementation of the NPT Safeguards Agreement of the Socialist People's Libyan Arab Jamahiriya', Report by the Director General, 20 February 2004, GOV/2004/12, http://www.iaea.org/Publications/Documents/Board/2004/gov2004-12.pdf.

8 IAEA, 'Implementation of the NPT Safeguards Agreement in the Islamic Republic of Iran', Report by the Director General, 10 November 2003, GOV/2003/75, http://www.iaea.org/Publications/Documents/Board/2003/gov2003-75.pdf.

9 IAEA, 'Implementation of the NPT Safeguards Agreement in the Republic of Korea', Report by the Director General, 11 November 2004, GOV/2004/84, para. 38, http://www.iaea.org/Publications/Documents/Board/2004/gov2004-84.pdf.

10 IAEA, 'Implementation of the NPT Safeguards Agreement in the Arab Republic of Egypt', Report by the Director General, 14 February 2005, GOV/2005/9, para. 22, available at http://www.carnegieendowment.org/static/npp/reports/gov2005-9.pdf.

11 IAEA, 'Implementation of the NPT Safeguards Agreement in the Republic of Korea', GOV/2004/84, para. 38 (emphasis added).

12 *Ibid.*, para. 41.

13 IAEA, 'Implementation of the NPT Safeguards Agreement in the Arab Republic of Egypt', GOV/2005/9, para. 22 (emphasis added).

14 *Ibid.*, para. 23 (emphasis added).

15 *IAEA Safeguards Glossary*, 2001 ed., International Nuclear Verification Series no. 3, section 2.2(d), http://www-pub.iaea.org/MTCD/publications/PDF/nvs-3-cd/Start.pdf.

16 *Ibid.*

17 If a so-called revised Code 3.1 provision has been agreed to under the Subsidiary Arrangements concluded with the agency.

18 *IAEA Safeguards Glossary*.

19 The Safeguards Implementation Report, submitted every year to the board, provides a description and analysis of the agency's safeguards operations during the previous year,

including a summary of the problems encountered and the secretariat's findings and conclusions. Distribution of the full report is restricted to IAEA member states and it is not available to the general public.

20 IAEA, 'Implementation of the NPT Safeguards Agreement of the Socialist People's Libyan Arab Jamahiriya', Resolution adopted by the Board on 10 March 2004, GOV/2004/18, http://www.iaea.org/Publications/Documents/Board/2004/gov2004-18.pdf; IAEA, 'Implementation of the NPT Safeguards Agreement in the Islamic Republic of Iran', Resolution adopted on 24 September 2005, GOV/2005/77, http://www.iaea.org/Publications/Documents/Board/2005/gov2005-77.pdf.

21 Article III.B.4 states that 'if in connection with the activities of the Agency there should arise questions that are within the competence of the Security Council, the Agency shall notify the Security Council, as the organ bearing the main responsibility for the maintenance of international peace and security'.

22 IAEA, 'Implementation of the NPT Safeguards Agreement in the Islamic Republic of Iran', GOV/2005/77.

23 The commitment not to produce feed material was important because without it Iran could not produce low- or highly enriched uranium domestically. However, in summer 2004 Iran notified the agency that it intended to resume enrichment-related activities and the production of UF_6 as feedstock.

24 'Statement by the Iranian Government and visiting EU Foreign Ministers', *In Focus: IAEA and Iran*, 21 October 2003, http://www.iaea.org/NewsCenter/Focus/IaeaIran/statement_iran21102003.shtml.

25 UN Security Council, 'Statement by the President of the Security Council', 22 April 2004, S/PRST/2004/10, available at http://www.unhcr.org/refworld/pdfid/411728c84.pdf.

26 Ann MacLachlan, 'France, Libya Initial Nuclear Pact, Emphasize Infrastructure-Building', *Nuclear Fuel*, vol. 48, no. 50, 13 December 2007, p. 15.

27 IAEA, 'Implementation of the NPT Safeguards Agreement in the Socialist People's Libyan Arab Jamahiriya', Report by the Director General, 12 September 2008, GOV/2008/39, http://www.iaea.org/Publications/Documents/Board/2008/gov2008-39.pdf (emphasis added). Such a conclusion raises some questions. First, the word 'continue' seems to imply that in the past the agency has implemented safeguards in Libya as a routine matter. This is not the case, since, as indicated in the report to the board, the agency had 'unrestricted and prompt access, *beyond* that required under its Safeguards Agreement and Additional Protocol, to those locations, information and individuals deemed necessary by the Agency to fulfil its verification requirements'. And secondly, why does the agency feel it necessary to officially state *now* that it will implement safeguards in Libya as a routine matter when the report acknowledges that there are still some unresolved issues such as the origin of 2 tonnes of UF_6 as well as other uranium compounds obtained

through a clandestine source, that 'there are some parts of Libya's past programme which the Agency has not been able to reconstruct fully' and that the agency still needs to 'work to reach a conclusion about the absence of undeclared nuclear material and activities in Libya'? The adoption by the board, on 24 September 2008, of a resolution supporting the continued implementation of safeguards in Libya 'as a routine matter' thus constitutes another detrimental precedent.

28 IAEA, 'IAEA Board Concludes Consideration of Safeguards in South Korea', staff report, 26 November 2004, http://www.iaea.org/NewsCenter/News/2004/south_korea.html (emphasis added).

29 *Ibid.*

30 This was done only on 24 September 2005 and it was not until 4 February 2006 that the board decided to report the matter to the UN Security Council.

31 Such a conclusion means that 'the Secretariat has found no indication that, in its judgement, would give rise to a possible proliferation concern'. IAEA, 'Safeguards Statement for 2007', para. 13, http://www.iaea.org/OurWork/SV/Safeguards/es2007.html.

32 At least ten AVLIS-related experiments involving exempted or undeclared nuclear material were carried out between 1993 and 2000 (IAEA, 'Implementation of the NPT Safeguards Agreement in the Republic of Korea', GOV/2004/84, para. 15). In 2002 and 2003 South Korea refused requests by the agency to visit KAERI's Laser Technology Center (para. 14); it refused to acknowledge in 1999 having conducted plutonium separation experiments (para. 26); and it did not report in August 2004 all past conversion activities (para. 20).

33 IAEA, 'Implementation of the NPT Safeguards Agreement in the Arab Republic of Egypt', GOV/2005/9 (emphasis added).

34 IAEA, *Annual Report for 2004* (Geneva: IAEA, 2005), p. 65, http://www.iaea.org/Publications/Reports/Anrep2004/safeguards.pdf.

35 The February 2005 report to the Board (IAEA, 'Implementation of the NPT Safeguards Agreement in the Arab Republic of Egypt', GOV/2005/9, para. 26) states 'The Director General will continue to report to the Board of Governors on the implementation of safeguards in Egypt as appropriate'.

36 *Ibid.*, para. 23.

37 IAEA, 'Implementation of the NPT Safeguards Agreement in the Arab Republic of Egypt', GOV/2005/9, para.24.

38 IAEA, 'Implementation of the NPT Safeguards Agreement in the Syrian Arab Republic', Report by the Director General, 19 November 2008, http://www.isis-online.org/publications/syria/IAEA_Report_Syria_19Nov2008.pdf.

39 Paragraph 19 of IAEA, 'The Structure and Content of Agreements Between the Agency and States Required in Connection with the Treaty on the Non-Proliferation of Nuclear Weapons', INFCIRC/153, provides that 'if the Board upon examination of relevant information reported to it by the Director General finds that the Agency is not able to verify that there has been no diversion of nuclear material ... to nuclear weapons or other

nuclear explosive device, it may make the reports provided for in paragraph C of Article XII of the Statute' including a report of non-compliance to the UN Security Council.

40 IAEA, 'Strengthening of Agency Safeguards: Special Inspections', 12 November 1991, GOV/2554 Attachment 1, para. 4.

41 Article 18 provides that if the board 'decides that an action by the State is essential and urgent in order to ensure verification that nuclear material ... is not diverted to nuclear weapons ... the Board shall be able to call upon the State to take the required actions without delay'.

42 IAEA, 'Excerpts from Statements made by the Director General under the Agenda item "Strengthening of Agency Safeguards" at the Board's December 1991 Meetings', 22 January 1992, GOV/INF/646, Attachment 1, para. 139.

43 IAEA, 'Strengthening of Agency Safeguards: Special Inspections', 12 November 1991, GOV/2554 Attachment 1, para. 14.

44 The secretariat concluded that all nuclear material in South Korea remained in peaceful activities after 'the Agency was able to clarify all issues relating to past undeclared activities'. IAEA, 'Safeguards Statement for 2007', para. 33. How the secretariat reached this important conclusion after the failures and breaches reported to the board in November 2004 (IAEA, 'Implementation of the NPT Safeguards Agreement in the Republic of Korea', GOV/2004/84) is described in Appendix I of the Safeguards Implementation Report for 2007. The latter is unfortunately not publicly available and should be made part of the reports transmitted to the Security Council.

45 Egypt should consider the merit of spontaneously agreeing to implement on a voluntary basis the Additional Protocol until such time the secretariat has concluded that Egypt's declarations are correct and complete and that there is no undeclared nuclear material and activities in Egypt. Such a commitment, while not departing from Egypt's principle decision not to sign and ratify at this stage the Additional Protocol, would make it clear that its principle decision is in no way intended to prevent the agency from drawing the necessary conclusion mentioned above, thereby distancing itself further from Iran's behaviour.

46 For instance, the Safeguards Implementation Report for 2000 reported that, by the end of that year, all states with facilities containing safeguarded nuclear material, except five which were named, had agreed to provide early design information on new facilities. One year later all had done so except Iran, which adapted its subsidiary arrangements accordingly in February 2003. The report for 2002 named four states which had not ratified their Additional Protocol more than five years after signature. Three of them did so within one year of the review of the report by the board.

47 IAEA General Conference resolution adopted on 4 October 2008, GC(52)/RES/13, http://www.iaea.org/About/Policy/GC/GC52/GC52Resolutions/English/gc52res-13_en.pdf.

Iran's Power in Context

Shahram Chubin

Iran–US relations – strained at the best of times since the 1979 Iranian revolution – have never been worse than during the past six years, due to the much more intense interaction between the two states since the revelations about Iran's nuclear ambitions and the United States' invasion of Iraq. The United States sees Iran as a potential strategic rival, while Tehran views the American presence in the Middle East as a potential existential threat. This has led to zero-sum thinking and has raised the stakes correspondingly. In the process there has been an inflation of the Iranian threat, which is poorly understood and often exaggerated. Depicting Iran as a military threat obscures the real *political* threat the country poses to its region; Iran's regional behaviour has been neglected and overshadowed by the contentious nuclear issue. However, it is precisely Iran's behaviour and goals which feed concerns about its nuclear ambitions.

It is important to put the Iranian threat in context. In recent years, what was largely a bilateral rivalry between Iran and the United States has become displaced and expanded throughout the region: Palestine, Syria, Lebanon, Iraq and the Gulf states have all been affected by the growing tension, and there are signs that Iranian influence is becoming stronger in these areas. Iran's more active and effective diplomacy in the Middle East is due to the conjunction of three separate trends, all of which are reversible. The first is the emergence of a permissive regional environment, hospitable to Iran's

Shahram Chubin is a Senior Non-resident Fellow of the Carnegie Endowment for International Peace. The opinions expressed in this article are his own.

Survival | vol. 51 no. 1 | February–March 2009 | pp 165–190 DOI 10.1080/00396330902749772

diplomacy; the second, the ascension of an ideological and hardline conservative government in Tehran, predisposed to a more activist diplomacy; and third, the oil windfall, which freed resources for seeding movements and clients supportive of Iranian goals. However, the influence that has accrued to Iran as a result of these trends is transitory and precarious, and there are constraints on Iran becoming a regional superpower.

Iran–US relations

A foundational myth of the Iranian revolution was that anti-Americanism was the equivalent of anti-imperialism.[1] This belief, together with the revolutionary desire to 'export the revolution', has become part of Iran's identity and *raison d'être,* which includes support for 'oppressed' people. Because these concepts are seen as vital to the regime's legitimacy, and hence security, they would be difficult to drop, not least because the sense of embattlement they engender is an integral part of maintaining elite control of the beleaguered and 'endangered' revolution.

In the Iranian view, the United States seeks to dominate the Middle East, and together with its local allies control the region strategically and loot its resources.[2] Iran's mission is to keep its independence by fighting this hegemonic and oppressive system and spreading its revolution by 'presenting a blueprint for an Islamic republic ... and defending the deprived masses in the world of Islam and the wronged people who have been trampled upon by tyranny'.[3] In the regional context this translates into calls for the expulsion of the Western presence and support for the Palestinian and other 'resistance' forces. Iran's challenge to the regional order is not military, but political: it is about exerting influence, appealing to the masses or the 'street', demonstrating that Iran can confront injustice, presenting a viable model of 'resistance', and 'framing the regional agenda'.[4] Thus, Iran is very far from being a 'classic imperial power' or 'an unexceptional opportunistic power seeking to exert preponderance in its immediate neighbourhood'.[5] Rather, it is a revisionist state in terms of status, not territory.

For the United States, the strategic importance of the Middle East has only grown in recent years, shaped by the confluence of such concerns as energy security, international terrorism and nuclear proliferation. Since the

Second World War, US policy toward the region has been encapsulated by the words 'access' and 'denial': Washington has sought political, military and economic access to the region and its resources through allies and bases, and to deny control over the region to any (unfriendly) dominant power. In recent years, the identity of the most likely rival power has shifted from the Soviet Union to Iraq and now to Iran.

In furtherance of its interests the US government has established bases in friendly states in the Persian Gulf (Kuwait and Qatar) and a long-standing naval presence (Bahrain). Washington envisages the evolution of an orderly security system as the Gulf States develop politically, establish competent military forces and cooperate collectively. Iran, with or without nuclear weapons, is seen as bent on the subversion of this system; its determination to 'extend its influence' throughout the broader Middle East is regarded as a 'grave threat to regional security',[6] as is its desire to exclude non-regional states. Iranian–American rivalry in the Persian Gulf, and beyond, is thus about two competing conceptions of regional order. The rivalry has only intensified as the two states have interacted at closer quarters.

Iran is seen as bent on subverting this system

Iranian defence officials say that Iran faces no threat from a regional state, hence the probable challenge is from 'an ultra-regional power like America'.[7] They argue that Washington is using the nuclear issue to limit Iran's regional power, which it sees as already so impressive that 'no country in the [Middle East] can stand against it'.[8] Although this tends to frame the potential development of an Iranian nuclear weapon as America's problem, the impact of such a development would be primarily regional, making the issue of direct concern to the Arab states, though they are reluctant parties to negotiations.[9] Similarly, no technical fix on the nuclear question will be viable unless the question of Iran's regional ambitions and role are addressed.[10]

Still, the idea that the Iranian–American rivalry – seen in its broadest sense as a battle between 'radical Islam' and the Western world – is at the core of Middle East politics is now widespread. The Palestinian issue, for example, has been reframed to focus on the threat from Hamas: said former Israeli Prime Minister Benjamin Netanyahu, 'any area we withdraw from will be taken over by Iran and its proxies'.[11] The 2006 war in Lebanon is seen by some as a war between Iran and Israel and others as a war between Iran and the United States.[12] There are those who argue the Annapolis conference in 2007 was motivated by Iran and the need to confront it.[13] Certainly, the enlarged US presence in the Middle East and the opportunities afforded by a changed regional context have allowed Iran to confront Washington and its allies on much wider canvas than before.

A permissive regional context

The Middle East has changed over the years in part because of some long-standing trends. Among these is the shift of the region's centre of gravity from the Levant to the Persian Gulf.[14] This has underscored divisions between the rich and poor members of the Arab world. Parallel to this has been the disappearance of Egypt as an Arab leader.[15] This has created a leadership vacuum, which accounts for the proliferation of regional initiatives in 2008 by other players. For example, Qatar has launched an initiative to address problems in Lebanon, Turkey in Syria, Egypt in Gaza, and Saudi Arabia in Afghanistan.

But American policies – and non-policies – have had the greatest effect on the regional landscape. US misjudgement and incompetence, exemplified by the country's entanglement and overextension in Iraq, is generally acknowledged. The effects on America's moral standing and credibility as a leader and ally have been devestating. Those elites in the region who believed in the United States have been discredited. Other policies, such as the failed 'freedom agenda' and the war in Lebanon, have fared little better. Neglect of the Palestine issue has allowed the 'two-state solution' to slip away as a realistic alternative for Israeli and Palestinians alike,[16] which has in turn fed extremism, rage and anti-Americanism.[17] The growing strength and appeal of the rejectionist front led by Iran is directly attributable to US policies.[18] Meanwhile, by shattering the Iraqi state, the United States eliminated Iraq as an effective regional buffer vis-à-vis Iran, whose influence over its neighbour immediately increased.[19] It meant that the old triangular system, in which the three large Gulf powers of Iran, Saudi Arabia and Iraq balanced one another, was replaced by a bipolar structure pitting Iran and Saudi Arabia directly against each other.[20] Moreover, by empowering Iraqi Shi'ites, the US occupation emboldened this traditionally suppressed community in the Gulf, a sensitive security issue for Saudi Arabia and even more so for Bahrain.[21] (This is not to say, however, that the Saudis see Iran primarily as a Shia threat – on the contrary, Iran is mostly seen as a rival for political power in the region.)

Questions about US judgement and reliability accelerated tendencies among the Gulf Co-operation Council (GCC) states to diversify their foreign policies and pursue relationships with other states.[22] It also weakened any impulse to follow Washington's lead, which explains why most Arab states have played a spectator role in Iraq, refusing to open embassies or forgive debts contracted under Saddam Hussein. This has effectively given Iran a free hand in Iraq.[23] If any Gulf state had been disinclined to confront Iran before the US invasion of Iraq, they were all the more diffident afterwards. Arguments for not confronting Tehran, such as a natural wish to accommodate a powerful state, appeared more convincing in the wake of questions about the reliability and judgement of the United States. Given their proximity to Iran, the GCC states had little desire to join any polarising coalition pitting Arab versus Persian.[24] Divisions within the Arab world also encouraged a cautious approach; while some states, such as Qatar, Oman and Dubai, enjoyed good relations with Iran, Saudi Arabia, Kuwait, Bahrain and Abu Dhabi sensed a security threat.[25] Finally, intimations that the United States might be willing to engage Iran diplomatically, as suggested by the Iraq Study Group report of 2006 (the Baker–Hamilton report) and the US National Intelligence Estimate (NIE) of November 2007, underscored suspicions among Arab countries of a possible US–Iran 'grand bargain' at their expense, encouraging them to take out further insurance against this eventuality.[26]

Indeed, the poorly worded 2007 NIE was the diplomatic equivalent of an 'own goal': it succeeded in undermining the emerging consensus on stronger measures against Tehran and virtually eliminated the military option as a diplomatic instrument. It may have emboldened Iran.[27] Certainly it undermined domestic critics of the regime's nuclear policies. US failures such as these have provided openings for Iranian diplomacy.[28]

None of this is to suggest that the Arab world is indifferent to the threat posed by a resurgent Iran seeking a nuclear capability. While governments may be reluctant to antagonise Iran – in some cases, their own populations exult in Iranian rejectionism – they have let their opinions be known indirectly. Arab newspaper editor Abd Al-Bari Atwan has suggested that Iran has benefited from the 'disastrous mistakes' of others without 'having to fire a bullet'.[29] Arab journalist Abd al Rahman Al- Rashid noted that Iran's missiles had a longer range than necessary if intended for Israel and concluded they must be intended 'to obtain special privileges'.[30] A third journalist, Tarik Al-Humayd, questioned whether Iranian support for Hamas and Hizbullah, and their designation as 'freedom armies', amounted to the 'hijacking [of] Arab causes to serve Iran's interests'.[31] Said writer Jameel Theyabi, 'Hezbollah is the solid arm of the Iranian regime in the region, while Hamas is about to become the second Iranian arm ... Syria too is working under the umbrella of Iran.'[32]

Iranian diplomacy in action

Many Iranians believe that their country has turned a corner and reached a new stage internationally. Commentators point to 'the gradual transfer of power and influence from America's camp to Iran's camp' and see the spread of Islamism in the region going hand-in-hand with 'the inclination of regional states to gravitate towards Iran'.[33] Hossein Shariatmadari, editor-in-chief of the Iranian newspaper *Keyhan* and a confidant of Iran's Supreme Leader Sayyid Ali Khamenei, has described the Middle East as a platform to demonstrate US 'failures and disappointments'.[34] Others have called for Iran to challenge the United States for leadership of the region through a 'proactive and aggressive foreign policy'.[35] The necessary condition for Iran's 'advancement' and greater freedom of action, in Ali Larijani's words, is the collapse of the America's 'exclusive hegemony while suffer[ing] a defeat'.[36]

A sense of growing power, new to Islamic Iran, has led to a certain braggadocio and recklessness and a coarsening of language that in fact does little to advance Iran's interests.[37] In what has been called the 'Palestinisation' of Iran's foreign policy,[38] Iran has insisted that Israel has no right to exist; that only one solution can work in Palestine, namely the one-state solution, whatever the Palestinians might decide; that the Holocaust is a myth; and that Iran's 'international mission' is to limit Israel's power.[39] In brief, Iranian President Mahmoud Ahmadinejad has made 'anti-Israeli rhetoric the defining characteristic of his presidency'.[40] Khamenei has blessed all of this, arguing that Iran has a duty to stand up to US bullying and 'the iniquities of the Zionist regime'.[41] Khamenei's international-affairs adviser, Ali Akbar Velayati, amplifies this thinking: 'to support Palestine is to support Iran ... To support Lebanon is to support Iran ... To support Iraq is to support Iran.'[42]

Iranian officials, believing that 'regional and global developments have created new situations in Iran's favour',[43] have made tireless efforts to cash in by spreading the message that the United States is 'hated in the region'.[44] But Iran's greater involvement in Middle Eastern affairs has not been merely opportunistic but also motivated by defence considerations. Though eager to spread its model, Iran was also impelled to extend its influence further afield by its sense of encirclement and the need to engage the enemy in 'forward defense, so as to avoid fighting on its borders. Palestine, in this view, is not a matter of solidarity but strategy: it is now a bulwark – a frontline – in Iran's defense'.[45]

Iran has filled a vacuum

Thus, in recent years Iran has increased its support of the Palestinian resistance. It has been helped in this project by two factors: firstly, the absence of any substantive peace process; and secondly, the fading prospects of a two-state solution.[46] This has meant that Iran has effectively filled a vacuum. What used to be provocative support for a 'one-state solution' now appears less extreme. When there is no peace process to oppose, looking at alternatives appears more reasonable.

The controversial enhanced US military presence in the region and Washington's extraordinary neglect of the Palestinian issue have provided fertile ground for Iran to pose as Palestine's protector and to generalise its conflict with the United States. The anger and frustration of many Arabs and Palestinians at the blatant disregard for their concerns have made some susceptible to extremist solutions. Iran's willingness to activate the rejectionist front, and to stand up to America, thus has regional resonance, which it would not have had in a different context or with different US policies.

Related to this is Iran's vaunted 'Hizbullah model' of resistance, which allegedly accounted for the success of that militia in holding off the powerful Israeli forces for 33 days in 2006. Ali Larijani, now speaker of Iran's parliament, has argued that the 2006 war punctured the myth of Israel's invincibility and demonstrated Iran's success in converting the Arab states to Iran's view. Said General Yahya Rahim Safavi, military adviser to the Supreme Leader, 'The combat model of the Palestinians has changed from a focus on Arab ethnical issues to an Islamic struggle under the banner of the Koran and Islamic slogans'.[47] Larijani claimed the Israeli defeat was 'an outcome of the Iranian Islamic revolution'. Ezatollah Ghafarizadeh, a senior Iranian military officer, claimed that Iran's own resistance was the model for Hizbullah's success by showing it was possible to defeat Israel. For Ahmadinejad the lesson was clear: 'resistance is the only way to defeat Israelis and their masters'.[48]

Iran's belief that negotiations will lead nowhere without 'resistance' has fuelled its support not just for Hizbullah but also, more recently, for Hamas.[49] Hamas accepts this reasoning: said a Hamas spokesman, 'what we learned from Hezbollah is that resistance is a choice that can work'.[50] If Hizbullah can eject Israel from Lebanon in 2000 and fight it to a standstill in 2006, there may be lessons for other resistance forces. And if Iran can claim to be the tutor of the Shia organisation Hizbullah, what better way to bridge the sectarian gap than to cross over to sponsorship of a Sunni resistance force like Hamas, especially when it has become a political party on a par with Fatah?

Iran's activism is not without its regional critics, however. Egypt's Foreign Minister Ahmed abu al-Gheit depicted Iran as 'on a surge', playing many cards in Iran's but not in the Arab world's interest: 'What we see in Lebanon [are] gains by an Islamic party by which I mean Iran'.[51] The head of Egypt's Foreign Committee was more blunt: 'We will not allow Iran to export the Islamic revolution to Egypt', he said, adding that Egypt would not tolerate the presence of an 'Iranian Islamic Emirate in Gaza'.[52]

Iran's only steadfast Arab ally since the revolution has been Syria. Originally, a shared antipathy to Saddam's Iraq was the cement holding this odd couple together, but a shared sense of being threatened by the

United States has caused the countries to become closer than at any time since the late 1980s.[53] Iran supplies arms to Hizbullah with Syrian assent, and the Syrian regime regards the 'Tehran–Damascus–*Hezbollah* axis' as the only thing keeping the United States and Israel from dominating the entire Middle East.[54] Neither state wants to see Iraq emerge as a US ally or base; both would prefer to see the United States exit humiliated and bloodied from their neighbour.

Indeed, since the demise of Saddam Hussein, Iran has largely seen Iraq 'through the lens of its conflict with the United States'.[55] The prime consideration for Iran is to prevent the emergence of an Iraqi threat. Tehran aims to do this by influencing the shape and orientation of a future Iraq, not by exerting territorial control.[56] Iran's foremost concern is the continuing US military presence, as Khamenei emphasised to Iraqi Prime Minister Nuri al-Maliki in mid 2008.[57] A permanent presence would mean 'control and domination', and former Iranian President Hashemi Rafsanjani pointedly warned that 'if the US remains in Iraq, there will be no security in Iraq or the region'.[58]

Iran's current concern is to ensure that any Status of Forces Agreement does not allow major US bases in Iraq and that any agreement specifically prohibits the use of Iraqi territory from attacks on neighbouring countries.[59] The December 2008 agreement between Iraq and the United States leaves Iran relatively reassured in this respect. Tehran expects under these circumstances to be able to influence the future course of Iraqi politics at least enough to assure its own security interests. However, its activism in Iraq and support for various Shia as well as other groupings has not gone unnoticed. The political empowerment of the Shia Muslims in Iraq and the example this poses for other similarly oppressed Shi'ites in the Gulf States (notably Bahrain) has increased suspicions about Iran's intentions amongst the country's Arab neighbours.[60]

Iran has shown insensitivity in other dealings with its Persian Gulf neighbours, further complicating an already fragile environment. Putting aside the suspicions engendered by the gaps between Sunni and Shia, Arabs and Persians and smaller Arab states and their larger neighbour, Iran's pretensions to Muslim leadership were bound to create strains. During Iran's attempts to 'export the revolution' in the 1980s, most directly to Bahrain and Kuwait, the Arab states combined to side with Iraq and oppose Iran.[61] The Gulf Co-operation Council was formed in 1981 with Saudi encouragement, and since then the GCC states have been wary of Iran, adopting different postures and mixes of deterrence and accommodation. All of these states to varying degrees look to the West, and primarily the United States, to offset and balance their powerful neighbour. None is interested in confronting Iran or otherwise provoking it, but equally, none is interested in supporting Iran's pet project: the expulsion of the United States from the Gulf and the substitution of a regional security arrangement through defence cooperation with Iran.[62]

Indicative of Iran's obsession with its own revolutionary principles is an obliviousness to others states' concerns and the impact of its utterances on those states' receptivity to Iranian initiatives. This is reflected in a tendency periodically to show a disregard for its neighbours which borders on contempt. For instance, in July 2007 Shariatmadari resurrected Iranian territorial claims to Bahrain. A year later, Deputy Foreign Minister Manouchehr Mohammadi suggested that 'the next crisis [for the Persian Gulf] is the legitimacy crisis of monarchical and traditional regimes. Considering today's conditions these systems cannot survive.'[63] Iran has made efforts to appeal to the Arab street over the heads of their governments: Khamenei has observed that there is an 'unfortunate' gap between the sentiments of the populace and the policies of their governments, saying 'they don't have the same voice as their nations'.[64] Shariatmadari has called for attacks on Israel, including its citizens worldwide, threatening that if Arab governments resisted they should be toppled.[65] Playing the 'Arab street' card may not be directly threatening to Iran's neighbours, but it puts them on notice about their potential vulnerability to Iranian propaganda.[66] At the same time, a decision to expand Iran's presence on the island of Abu Musa, long disputed with Abu Dhabi, managed to unite the otherwise fractious Arabs against this 'illegitimate action on an indivisible part of the UAE'.[67] The Gulf states might be forgiven, in light of all this, for assuming that the Iranian offer of a defence pact resembles a Chicago-style protection racket.

A hardline president

Security policy in Iran is the product of a 'consensus' formed in the Supreme National Security Council. Though presided over by the president, in reality the final decisions are made by the Supreme Leader. In his willingness to do anything to preserve his position, Khamenei is decidedly pragmatic, but by inclination and preference he is a hardliner, comfortable with pure revolutionary ideologues like Ahmadinejad. Ahmadinejad represents a new generation of leader, forged in the war with Iraq and dedicated to the revolution's ideals (promoting the Islamic model, seeking social justice, fighting oppression), which they see as having been betrayed by the previous president, Mohammad Khatami, who pursued a policy of fence-mending and détente toward Iran's Persian Gulf neighbours and Europe.

Ahmadinejad is a passionate ideologue and populist demagogue who has enlisted a new, poor, rural constituency and helped revive the flagging revolutionary spirit. He also has a special constituency, the Revolutionary Guard, whom he has rewarded accordingly.[68] He respects and defers to the Supreme Leader, who finds his firebrand style convenient to sideline the reformists, who threatened to make his post obsolete. Khamenei has endorsed Ahmadinejad's confrontational style, to the point where the Supreme Leader – uncharacteristically – is identified with it, and thus with a particular faction within the Iranian government.[69] With the perception that the United States is 'on the run' there is little incentive for him to depart from his own preferences and rein in his favourite president.

Ahmadinejad is a passionate ideologue and populist demagogue

In Ahmadinejad's narrative of embattlement, the whole world has 'ganged up' on Iran, which valiantly stands for principle and moral values.[70] This 'Cuban option' has its uses, excusing failures and disappointments by reference to the actions, such as international sanctions, of others.[71] Ahmadinejad uses confrontation to enhance his domestic standing.[72] By tapping into Iranian national pride and nationalism he has tried to energise support for the regime and the revolution. And by depicting his predecessors as faint-hearted, willing to sell out the revolution's aspirations and principles, he has cornered the market on an activist, offensive foreign policy. In this he has the support of Khamenei: Iran, says Khamenei, has 'an offensive stance (*Tahajomi*) ... If ever we can deliver a political blow to America or Israel, we will surely do so.'[73] This 'dynamic foreign policy', according to Khamenei, is more in line with 'revolutionary concepts' than the mistaken 'passive policy' of the previous government and of Ahmadinejad's contemporary critics: 'Some want to undermine the revolution by using the pretext of rationality, moderation and refraining from turmoil. This cannot be done ... It could be that they are tired themselves but attribute it to the people.'[74]

Ahmadinejad's confrontational policies have been congenial to the leadership and so far relatively free of cost. By mobilising the masses and appealing to the Arab street, the president has given the Supreme Leader wider recognition and popularity in the Arab world and reinvigorated his revolutionary core of supporters within Iran.[75] By marginalising the reformists and discrediting their passive, reactive policies as little more than subservience, Khamenei has made criticism of Ahmadinejad's policies more difficult. US mistakes and failures, including the ill-advised NIE, have also made it harder for critics to demonstrate the risks and costs of Ahmadinejad's policies. But this could change: a major setback in the region that marginalised Iran or caused a united Arab front to emerge; a breakthrough agreement by Hamas in Palestine or Hizbullah in Lebanon; or a clash with the United States could ignite controversy and intensified debate in Iran about the wisdom of the prevailing course.

Oil: a two-edged sword

Iran's regional activism and support of the Hamas and Hizbullah 'resistance' movements have been aided by the oil windfall of recent years. But as the term 'oil curse' suggests, a fall in the price of oil can succeed its rise with dizzying rapidity, and can do so after recipients of oil rents have become addicted or at least habituated to them. Iran has additional problems: it has a record of economic mismanagement and is under international sanctions. The recent decline in oil prices threatens to have a major impact on Iran's economy and ability to continue as paymaster of the 'resistance', and on Ahmadinejad's political fortunes.

According to the Central Bank of Iran, oil generates 80% of foreign earnings and 60% of government revenues, or roughly 30% of GDP in a given year, although this varies according to oil prices.[76] Whatever the exact statistics, it enough to note that in the first half of 2008 Iran received $54 billion in oil revenues, which is double what Iran was earning per year during the Khatami era.[77] When oil prices were high, Iran had plenty of cash to spread around as it pursued its more offensive foreign policy. Exact figures are elusive, not least because cash payments were complemented by arms, training and contributions to social or charitable activities, but it is clear that Iran has gone beyond the usually cited $100 million per year payment to Hizbullah in recent years. For example, in 2006, in addition to its usual contribution, Iran re-supplied Hizbullah with arms consumed in that summer's war with Israel.[78] Iran also paid $10,000, via Hizbullah, to every Lebanese citizen whose house had been destroyed in the war. One source puts Iran's total disbursement to Hizbullah in 2006 at $1.2bn.[79] In addition, Ahmadinejad made commitments to Islamic Jihad and Hamas, and the Iraqi militias Jaish al-Mahdi, the Badr Brigade and other Sunni equivalents. In July 2005 Iran offered the Iraqi government $1bn in aid.[80]

It is impossible to put an accurate price tag on the annual costs of all these commitments. It is clear, however, that these investments in expanding Iran's regional influence have grown with the oil windfall and will contract with the oil-price slide. The gap between $147 oil in the first half of 2008 and $50 in the last half is bound to curtail Iran's capabilities, if not its ambitions.[81] Indeed, falling oil prices will almost certainly have painful domestic consequences as well. The boom in oil revenues increased dependence on the commodity, postponed necessary reforms and cushioned Iran from the effects of international sanctions.[82] Economic mismanagement, which has fuelled inflation and popular expectations, will be exposed for what it is with sharply diminished revenues and the pessimism engendered by a global recession that is likely to last several years.[83]

If oil prices continue to slide, Iran will soon face some unattractive choices. The government has already raided the oil stabilisation fund. According to some sources, over the last three years Iran's foreign-currency reserves have been regularly siphoned off to current expenditures, leaving Iran with anticipated reserves of less than $10bn by the end of 2008.[84] This implies a growing budgetary deficit, which according to the International Monetary Fund is unsustainable if oil prices remain in their current range under $75, as Iran has 'limited access to international financial markets due to sanctions'.[85] Given its shortfall in revenues, if the government is unwilling to cut back on expenses it will be faced with a choice between printing money (an inflationary strategy) or seeking to add to revenues by raising taxes or reducing subsidies. Its attempt to impose a relatively modest sales tax of 3% this year brought the bazaar out on strike and was quickly reversed.[86] Cutting subsidies will be no more attractive for a populist president. If, as seems likely, demand for oil remains limited, reflecting the sluggish global economy, Iran cannot count on an increase in oil revenues in the near term. It is an ugly reminder to Iran of its limitations as a producer: having no spare capacity, it cannot look forward to benefiting from increased production, whether prices rise again or not.

Iran cannot count on an increase in oil revenues in the near term

Mismanagement and the slowdown will make Iran's economy more vulnerable to sanctions, which have already begun to bite in the financial sector. UN Security Council sanctions leave Iran in a poor position to seek access to international credit. Sanctions have already crimped Iran's trade with the United Arab Emirates, principally Dubai.[87] Further, the country has been unable to invest in the maintenance and upgrading of its oil infrastructure and has difficulty maintaining its current production rate of nearly 4m b/d. With inadequate refining capacity for its domestic needs Iran would be very vulnerable to sanctions that targeted its imports of petroleum products, especially gasoline.[88] Unfortunately for Tehran, the prospect of more sanctions is increased by the election of US President Barack Obama. The new president will come with none of the baggage of his predecessor and a greater willingness to engage in diplomacy, but with no greater tolerance for Iran's spoiler policies in the region or its nuclear ambitions.

Iran will not collapse because of decreased revenues; it survived a similar situation in the 1980s and will, no doubt, try to make a virtue out of necessity by appealing to the revolutionary spirit while blaming sanctions and foreign enemies for any difficulties. It is likely, however, that Ahmadinejad's policies of confronting the international community, supporting resistance forces, throwing money at problems against specialist advice and making populist promises of improved living standards will face increased domestic criticism. The opportunity costs of funding insurgents and militias may become more controversial.[89] After all, by all accounts the Iranian street is no fan of these policies. Some believe that economic constraints will hamper Iran's regional ambitions.[90] Ahmadinejad's economic and political model may be discredited, especially if a new round of meaningful sanctions is imposed in the first half of 2009. At the very least, economic constraints should encourage a domestic debate about what kind of regional role Iran should play, and at what cost.

Constraints and limits to Iranian power

Iran has a great past and potentially a great future, but has been stuck in a revolutionary rut for three decades. As a result, it has exerted itself only to counter perceived threats to its revolutionary identity and to promote its mission abroad. Tehran has become habituated to seeing threats everywhere: in the West's presence in the Middle East; in the existence of Israel (surely a discretionary foe?); and in the forces of secularism, materialism and globalisation ('cultural threats'), to name a few. Meanwhile, the real threat to the country's security – instability in Iraq and Afghanistan – has been treated as a battleground in the fight against the United States. An embattlement mindset suits the ideologues of the regime who benefit from the country's beleaguered status and prevent it from becoming a normal, independent and respected state. Their emphasis on Iran's rights is not matched by discussion of its responsibilities; their insistence on a narrative of victimhood is not balanced by respect for others' concerns.

Iran is not a marginal state like Libya or Syria, but neither is it a great power. It has few friends and fewer allies. By alienating the United States and Europe, Iran has increased its dependence on Russia for diplomatic support, nuclear and other technology and conventional arms. It has thus compromised its vaunted 'independence', inhibiting the pursuit of its interests in the Caspian and Caucasus. Iran has neither hard nor soft power. It has leveraged US mistakes in the last few years as its principal source of influence in the region. Playing on regional frustrations and anger, Iran has positioned itself as a spoiler and given rejectionism a new fillip in the Arab street. But these are limited and wasting assets, dependent on continued US errors and the failure of peaceful alternatives for Palestine. Iran can offer a rejectionist 'war option' but not a solution; for that the Arabs must turn to others. Iran's conventional military capabilities, especially its power-projection capabilities, are limited, even with respect to, or in comparison with, its immediate neighbours.[91]

Iran's conventional military capabilities are limited

Iran's economic model is a clear failure in comparison to that of its smaller southern neighbours, whence large numbers of Iranian entrepreneurs have fled. Unable to diversify its economy or invest in its energy infrastructure, Iran is storing up trouble for itself. With oil prices sliding Tehran has to rely on Saudi Arabia for cuts in production to protect it against further price deterioration and associated political repercussions.[92] The damaging effects of corruption, subsidies, a bloated government and the operations of *bonyads* (unaccountable 'foundations' that dominate the non-petroleum sectors) have caused the country to haemorrhage talent abroad.

Iran's political model has few if any takers, as even Iraq is demonstrating. A revolutionary government presides over a post-revolutionary society; the gap between state and society has widened. Iranian people – though not the government – may be said to have found the 'middle ground between quietism and martyrology'.[93] Nationalism alongside Islam is a part of a proud Iranian identity, but there is little appetite, or precedent, for regional domination or even leadership.

Iran's regional influence has come from a convergence of several factors: US mistakes and vulnerability, regional anger and the appearance of an ideological government in Tehran with access to unprecedented resources.

None of these are permanent. Iran's resources have already shrunk, the conservative government is being discredited and the United States is in the process of rehabilitation. The structural constraints on Iran have become more evident. The Arab Sunni states, though divided and ambivalent, are all concerned by Iran's forward policy and nuclear ambitions. None of them wishes to exchange the United States for Iran as a security manager of the region. The limits to Iranian influence are also clear. Hizbullah, as a militia, may be dependent on Iran for arms and training but is also a Lebanese political party. It is thus a partner rather than client. The same applies to Iran's relations with Syria, its sole Arab friend. In troubled times this relationship appears solid, but with a more flexible attitude on the part of Israel and the United States, Syria may drift away from Tehran's embrace.

In Iraq, Tehran has considerable influence. This is normal given the number and complexity of ties between the two neighbours. But even here, too much Iranian interference or overreaching risks a backlash. Favouring a particular militia in Iraq risks alienating others and creating blowback, for once they become strong, militias cannot easily be controlled. Even the vaunted region-wide ascendancy of the Shi'ites, which *appears* to favour Iran, in fact reveals the constraints on any putative leadership role for Tehran. Sectarian polarisation, whether in Iraq or throughout the wider region, would limit Iran to (at most) a sectarian constituency – a minority in the region – and constitute a setback to Iran's Arab street strategy designed to transcend the sectarian divide. Moreover, any emphasis on sectarian issues would range the Arab states defensively against Iran and bury any Iranian regional project in the Persian Gulf. Iran's perceived arrogance and attitude of 'blatant superiority', which treats the Arab states as 'hostages', is widely felt and resented, hampering any Iranian leadership potential.[94] In short, Iran remains a limited threat. Hyping Iran's power and importance encourages its parasitical policy of feeding on disorder and understates its vulnerabilities.[95]

The arrival of a new American president at last permits a policy that puts Iran into proper perspective. Too often in recent years the US discussion of Iran has in fact has been a debate about America itself, filled with partisan rancour and cant. Now it is time to see Iran for what it is: a political threat and regional nuisance with many weak points, which can be played on.

Neither a colossus nor a failed state, its primary focus is regime survival. It is now the job of the Western foreign-policy community to communicate firmly but quietly that regime survival and regional destabilisation may be incompatible.

Notes

1 See Olivier Roy, *The Politics of Chaos in the Middle East* (New York: Columbia University Press, 2008), p. 130. Anti-imperialism in practice became anti-Americanism, a slant that is all the more extraordinary given the record of Russian/Soviet predatory behaviour in the region.

2 Ayatollah Ali Khamenei, for example, expressed this view on Iran TV Channel on 8 May 2008.

3 Ayatollah Ali Khamenei, in a broadcast on Iran TV News Channel on 3 June 2008.

4 See Emil Hokayem, 'To Counter Iranian Ambitions, the Gulf Needs Greater Self-Confidence', *The National*, 14 August 2008, http://www.thenational.ae/article/20080814/OPINION/152591015/1002-36k-; and Shibley Telhami, 'It's Not About Iran', *Washington Post*, 14 January 2008, p. A21.

5 Richard Haass, 'The New Middle East', *Foreign Affairs*, vol. 85, no. 6, November–December 2006, p. 6; and Ray Takeyh and Vali Nasr, 'Getting Iran Inside the Tent', *International Herald Tribune*, 7 December 2007, p. 8. This exact formulation is repeated in Ray Takeyh and Mark Brzezinski, 'The Time For Threats and Insults is Past', *International Herald Tribune*, 12–13 January 2008, p. 4. For a gross exaggeration of Iran's power and importance see Robert Baer, *The Devil We Know: Dealing with the New Iranian Superpower* (New York: Crown, 2008).

6 Condoleezza Rice, 'Rethinking the National Interest', *Foreign Affairs*, vol. 87, no. 4, July–August 2008, p. 18. See also Thom Shanker, 'Gates Sees Iran As Still-Serious Threat', *New York Times*, 9 December 2007, http://www.nytimes.com/2007/12/09/world/middleeast/09gates.html.

7 Mohhamad-Bager Zolgadr, deputy chief of the Armed Forces, quoted in the Iranian newspaper *Farhang-e Ashti*, 26 August 2008, BBC Monitoring, 1 September 2008.

8 This claim has been made repeatedly by Ali Larijani, secretary of Iran's Supreme National Security Council 2005–07, and quoted in various sources: Islamic Republic News Agency website, 28 September 2005, BBC Monitoring, 29 September 2005; *Aftab-e Yazd*, 7 July 2008, BBC Monitoring, 9 July 2008; Islamic Republic News Agency website, 11 October 2008, BBC Monitoring, 13 October 2008; and Ali Larijani, Islamic Republic News Agency website, 26 June 2008, BBC Monitoring, 27 June 2008. Mohsen Rezai, secretary of Iran's Expediency Council, and others have made similar observations.

9 Jihad al-Khazin, 'Religious Sermon', *Al Hayat* website, 25 September 2008,

BBC Monitoring, 29 September 2008; Joseph Kechichian, 'Can Conservative Arab Gulf States Endure a Fourth War in the Persian Gulf?', *The Middle East Journal*, vol. 61, no. 2, Spring 2007, p. 303.

10 Shahram Chubin, *Iran's Nuclear Ambitions* (Washington DC: Carnegie Endowment for International Peace, 2006); Emily Landau, 'Now Is the Time to Talk', *Haaretz.com*, 15 July 2008, http://www.haaretz.com/hasen/spages/1001243.html.

11 Interview of Benjamin Netanyahu by Tobias Buck and Lionel Barber in *Financial Times*, 7 October 2008, p. 7.

12 Ze'ev Schiff, 'Israel's War with Iran', *Foreign Affairs*, vol. 85, no. 6, November–December 2006, pp. 23, 25–6. Richard Norton, 'Why Hizbullah is Winning', *Middle East Journal*, vol. 61, no. 1, Winter 2007, p. 147.

13 Saul Singer, 'Whatever Palestinians Discuss, The Real Issue is Iran', *Daily Star* (Beirut), 20 June 2008; and David Brooks, 'Present at the Creation', *New York Times*, 6 November 2007, www.nytimes.com/2007/11/06/opinion/06brooks.html.

14 Emile El-Hokayem and Matteo Legrenzi, 'The Arab Gulf States in the Shadow of Iranian Nuclear Challenge', Stimson Center Working Paper, 26 May 2006, p. 8; Marina Ottoway and Mohammed Herzallah, 'The New Arab Diplomacy: Not With the US and Not Against It', Carnegie Papers, No. 94, Middle East Program, July 2008, p. 15; and Olivier Roy, *The Politics of Chaos in the Middle East*, p. 252.

15 For a discussion of Egypt's ailing government and the Saudis' vain efforts to fill this vacuum, see 'Saudi Arabia: Can it Make Peace in the Wider Region?', *Economist*, 11 October 2008, p. 51.

16 Sari Nusseibeh, 'The One State Solution', *Newsweek*, 20 September 2008, http://www.newsweek.com/id/160030; Isabel Kershner, 'Memo From Jerusalem: Support For 2-State Solution Erodes', *New York Times*, 3 September 2008, http://www.nytimes.com/2008/09/04/world/middleeast/04state.html; Hussein Agha and Robert Malley, 'Into The Lion's Den', *New York Review of Books*, 1 May 2008, pp. 57–60; Olivier Roy, *The Politics of Chaos in the Middle East*, p. 140; Marina Ottoway, Nathan Brown, Amr Hamzavi, Karim Sadjadpour and Paul Salem, *The New Middle East* (Washington DC: Carnegie Endowment for International Peace, 2008), pp. 15, 17.

17 For an overview see Ottoway and Herzallah, 'The New Arab Diplomacy'.

18 See 'Coalition of the Unwilling', *Economist*, Special Report: The Arab World, 19 October 2006, p. 24. Ottoway et al., *The New Middle East*, p. 30; Rami Khouri, 'Why Everyone is Negotiating in The Middle East', Agence Global, 21 July 2008, http://belfercenter.ksg.harvard.edu/publication/18449/why_everyone_is_negotiating_in_the_middle_east.html.

19 Olivier Roy refers to the end of a Sunni Iraq as 'Shattering the Middle East Balance of Power', *The Politics of Chaos in the Middle East*, p. 108.

20 Henner Fertig, 'Conflict and Cooperation in the Persian Gulf: The Interregional Order and US Policy',

Middle East Journal vol. 61, no. 4, Fall 2007, p. 633.

21 See Henner Fertig, 'Conflict and Cooperation', p. 639; and Vali Nasr, 'When the Shi'ites Rise', *Foreign Affairs*, vol. 85, no. 4, July–August 2006.

22 This trend has been most marked in the case of Saudi Arabia, which seeks greater independence in foreign policy. Ottoway and Herzallah, 'The New Arab Diplomacy', p. 10.

23 See Karen de Young, 'Iraq Finds its Arab Neighbors Are Reluctant to Offer Embrace', *Washington Post*, 16 May 2008, p. 109.

24 For a good discussion of the 'structural' aspects of GCC–Iran relations see El-Hokayem and Legrenzi, 'The Arab Gulf States in the Shadow of Iranian Nuclear Challenge'; and Ottoway and Herzallah, 'The New Arab Diplomacy'.

25 With Iraq taken out of the balance, some states are also reluctant to entrust their future to Saudi Arabia.

26 On Arab suspicions regarding a possible grand bargain, see Fouad Ajami, *The Foreigners Gift: The Americans, the Arabs, and the Iraqis in Iraq* (New York: Free Press, 2007), pp. 213–14; and El-Hokayem and Legrenzi, 'The Arab Gulf States in the Shadow of Iranian Nuclear Challenge'.

27 'Nuclear Fallout', *Economist*, 8 December 2007, p. 57.

28 Ottoway et al., *The New Middle East*, pp. 35–7.

29 Abd Al-Bari Atwan, 'Arabs Urged to Build Strong Military to Counter Iran', *Al Quds Al Arabi* website, 27 September 2008, BBC Monitoring, 29 September 2008.

30 Abd al Rahman Al-Rashid, 'Iranian Threat, Now it is Europe's Turn', *Asharq Al-Awsat*, 10 July 2008, BBC Monitoring, 14 July 2008.

31 Tarik Al-Humayd, 'The Iranian Freedom Armies', *Asharq Al-Awsat* website, 28 October 2008, BBC Monitoring, 30 October 2008.

32 Jameel Theyabi, 'Arabs Between Washington and Tehran', *Al-Hayat*, 14 July 2008, http://english.daralhayat.com/opinion/OPED/07-2008/Article-20080714-2215194b-c0a8-10ed-0007-ae6d646c153a/story.html.

33 Payman Tajrishi (ed.), Iran website, 15 December 2007, BBC Monitoring, 16 December 2007; Hanif Ghaffari, Resalat website, 20 February 2008, BBC Monitoring, 25 February 2008.

34 *Shariatmadari Keyhan* website, 2 April 2008, BBC Monitoring, 4 April 2008.

35 Resalat website, 21 February 2008, BBC Monitoring, 26 February 2008.

36 Larijani Fars News Agency, quoted in *MEMRI*, no. 426, 12 March 2008.

37 For a discussion of Iran's risk-taking, see Shahram Chubin, 'Iran's Risk-Taking in Perspective', IFRI Proliferation Papers, Winter 2008. With respect to the coarsening of Iran's language, it is worth noting Larijani's comment that the United States should attack Iran 'only if it wanted to see Israel in a wheelchair'. Najmeh Bozorgmehr and Roula Khalaf, 'Dismay as Top Nuclear Official quits', *Financial Times*, 22 October 2007, p. 5.

38 The term 'Palestinisation' was coined by Jahangir Amuzegar in 'The Ahmadinejad Presidency: Preparing for the Apocalypse', *Journal of International Affairs*, vol. 60, no. 2, Spring–Summer 2007, p. 49.

39 'Ahmadinejad: Iran's "International Mission" to Limit Israeli Power', Deutsche-Presse-Agentur, 4 October 2007, http://www.iranfocus.com/modules/news/article.php?storyid=12677.

40 See David Menashri, 'Iran's Regional Policy: Between Radicalism and Pragmatism', *Journal of International Affairs*, vol. 60, no. 2, Spring–Summer 2007, pp. 153–67.

41 See 'Leader's Message on Condemnation of the Zionist Regime's Crimes Lebanon', Center for Preserving and Publishing the Works of Grand Ayatollah Sayyid Ali Khamenei, 1 August 2006, http://english.khamenei.ir//index.php?option=com_content&task=view&id=218&Itemid=26.

42 Cited in Amin Tarzi, 'The World's Ninth Nuclear Power: Iran's Ambitions in the Middle East and Beyond', *Turkish Policy Quarterly*, vol. 1, no. 2, p. 63.

43 Said Jalili, secretary of the Supreme National Security Council, Islamic Republic News Agency website, 15 October 2008, BBC Monitoring, 16 October 2008.

44 Ali Larijani, Islamic Republic News Agency website, 24 September 2008, BBC Monitoring, 25 September 2008; and Islamic Republic News Agency website, 15 October 2008, BBC Monitoring,17 October 2008.

45 See Delphine Minoui's interview of Ahmadinejad's adviser and intimate Mojtaba Rahmandoust in *Le Figaro*, 29 August 2006, p. 3. For a similar view of Iran's policy in Lebanon, see the well-informed Iranian commentator Sadegh Zibakalam, *Daily Star* (Beirut), 17 June 2008.

46 Michael Slackman, 'Arab Leaders Grow Doubtful of a 2-state Solution', *International Herald Tribune*, 23–4 February 2008, p. 3; Jon Kimche, 'Mideast Peacemaking: Keep Hope Alive', *International Herald Tribune*, 18 November 2008, p. 11.

47 Yahya Rahim-Safavi, military adviser to the supreme leader, Fars News Agency website, 26 September 2008, BBC Monitoring, 29 September 2008. See also Mohammad Reza Djalili, 'La Politique Arabe de l'Iran', *Contrario*, vol. 5, no. 2, 2008, p. 136.

48 See Ali Larijani, Peygham-e Emrooz, 6 February 2008, BBC Monitoring, 12 February 2008; Brig.-Gen. Dr Ezatollah Ghafarizadeh, Head of Ground Forces Office of Research and Strategic Studies, *Keyhan* website, 23 April 2008, BBC Monitoring, 28 April 2008; Hizbullah News website, 3 February 2008, BBC Monitoring, 5 February 2008; Mahmoud Ahmadinejad, Islamic Republic News Agency website, 10 March 2008, BBC Monitoring, 11 March 2008.

49 Hashemi Rafsanjani, interview with al-Jazeera, 28 June 2008, BBC Monitoring, 1 July 2008.

50 Quoted in Taghreed El-Khodary and Isabel Kershner, 'As Israelis Pull Out of Gaza, Hamas Celebrates With its Rocketry', *New York Times*, 4 March 2008, http://www.nytimes.com/2008/03/04/world/middleeast/04mideast.html?ref=middleeast.

51 'Egypt Sees Iran "On a Surge" in Arab World', *BusinessIntelligence Middle East*, 17 June 2008, http://www.bi-me.com/main.php?id=21421&t=1.

52 Asriran.com, 11 Aban 1387, quoted in *Iran Early Bird*, 2 November 2008.

53 See Fred Lawson, 'Syria's Relations with Iran: Managing the Dilemmas of Alliance', *Middle East Journal*, vol. 61, no. 1, Winter 2007, pp. 41–2.

54 Patrick Seale, 'The Syrian–Lebanese Embrace', *International Herald Tribune*, 24 October 2008, p. 8.

55 Vali Nasr, 'In Tehran's Setbacks, an Opportunity in Iraq', *Washington Post*, 19 June 2008, p. A19.

56 *Ibid.*

57 Sayyid Ali Khamenei, *Voice of IRI*, 9 June 2008, BBC Monitoring, 10 June 2008. See also Nazila Fathi and Richard Oppel, Jr, 'US Troops Causing Instability: Iran's Religious Leader Tells Iraqi Premier', *New York Times*, 10 June 2008; and 'A Partnership With Iraq', *Washington Post*, 15 June 2008, p. B06.

58 Hashemi Rafsanjani, interview with al-Jazeera, 28 June 2008, BBC Monitoring, 1 July 2008.

59 Ali Larijani, IRI News Network, 12 September 2007, BBC Monitoring, 13 September 2007. See also 'Iraq Wants US Curbed in Attacking its Neighbors', *International Herald Tribune*, 30 October 2008, p. 6.

60 See Vali Nasr, *The Shia Revival: How Conflicts Within Islam Will Shape the Future* (New York: Norton, 2006). For a more sceptical look at the Shia impact on regional politics, see Michael Broning, 'The Myth of the Shi'ite Crescent', *Daily Star* (Beirut), 21 May 2008; Pat Proctor, 'The Mythical Shia Crescent', *Parameters*, vol. 38, no. 1, Spring 2008, pp. 30–43; and Roger Shanahan, 'Bad Moon Not Rising: The Myth of the Shi'a Crescent', Lowy Institute Analysis, September 2008, available at http://www.lowyinstitute.org/Publication.asp?pid=893.

61 This has since coloured Arab perceptions of Iran as a revolutionary state. See Fertig, 'Conflict and Cooperation in the Persian Gulf', p. 629.

62 For a discussion of Iranian opposition to US bases in the Middle East (Kuwait, Bahrain, Oman, Saudi Arabia and Qatar), see Hashemi Rafsanjani, interview with *Al Jazeera*, 28 June 2008, BBC Monitoring, 1 July 2008. For a discussion of Iran's idea of defence cooperation see Defence Minister Mostafa Mohammad Najjar, 'We Consider Our Neighbours as Our Allies', Islamic Republic News Agency, 29 August 2008, http://www2.Islamic Republic News Agency.ir/fa/news/view/line-17/0808292026183234.htm. Former Iranian Revolutionary Guard Commander Yahya Rahim Safavi has reassured the GCC that Iran 'is not covetous of their lands and is even prepared to sign a joint defence pact with them', ISNA website, 23 May 2008, BBC Monitoring, 26 May 2008.

63 *Keyhan*, 9 July 2007. The GCC saw these statement as 'irresponsible ... grave ... hostile'. See Al Arabiya TV, Dubai, 7 August 2008, BBC Monitoring, 8 August 2008. For criticism of these statements by Iranian reformists, see *Mardom Salari* website, 9 August 2008, BBC Monitoring, 15 August 2008.

64 Sayyid Ali Khamenei, Iran TV news channel, 3 June 2008, BBC Monitoring, 4 June 2008. It is regrettable, he says, that some countries subordinate the interests of the region to those of the United States. See *Voice of IRI*, 8 February 2008, BBC Monitoring, 12 February 2008.

65 See Report of Mehr News Agency, 4 October 2007, BBC Monitoring, 5 October 2007. The more pointed threats came in an editorial in *Keyhan*, 26 January 2008, quoted in *MEMRI*, no. 1828, 27 January 2008.

66 Iran has an Arabic-language TV network, Al-Alam, aimed at an Arab audience.

67 GCC Secretary General Abdurrahman al-Attiyah, quoted in 'GCC Chides Iran Over Facilities on UAE-claimed Island', AFP, 16 August 2008, http://afp.google.com/article/ALeqM5jKWFwdIofnq7X_3btkzLU-y7tfwQ. See also Nicole Stracke, 'Where is the UAE Island Dispute Heading?', Gulf Research Center, 25 September 2008, http://www.grc.ae/?frm_action=view_newsletter_web&sec_code=grcanalysis&frm_module=contents&show_web_list_link=1&int_content_id=54549&PHPSESSID=5bf9fe1d73e0b01a40d421dfcce31333. For background detail on the Gulf states' differing responses to Iranian policies, see Dalia Kaye and Frederic Wehrey, 'A Nuclear Iran: The Reactions of the Neighbours', *Survival*, vol. 49, no. 2, Summer 2007, pp. 111–28.

68 See Akbar Ganji, 'The Latter-day Sultan', *Foreign Affairs*, vol. 87, no. 6, November–December 2008, pp. 45–66.

69 Although there is a broad agreement on the necessity for 'regime security', there are very different notions within the Iranian regime of how this is to be achieved and, indeed, how and in what direction the regime should evolve. Thus foreign policy in practice can vary from Khatami's relatively moderate policies to Ahmadinejad's confrontational approach. Khatami is not alone in *not* seeing Middle East politics in zero-sum terms. Rafsanjani and Mohammad Qalibaf (mayor of Tehran and probable presidential candidate) have criticised this approach as applied to Iraq and elsewhere, noting areas of common, overlapping interests with the United States and others. Mehr News Agency, 12 February 2008, BBC Monitoring, 13 February 2008; Scott Macleod and Nahid Siamdoust, 'A Gentler Iran', *Time*, 19 March 2008, http://www.time.com/time/magazine/article/0,9171,1723711,00.html.

70 Confrontational policies clearly serve factional ends as they tend to strengthen hardliners; see Sanam Vakil, 'Tehran Gambles to Survive', *Current History*, December 2007, especially p. 420.

71 See Tim Guldimann, 'The Iranian Nuclear Impasse', *Survival*, vol. 49, no. 3, Autumn 2007, pp. 169, 175.

72 Roy, *The Politics of Chaos in the Middle East*, pp. 126–7; see also Guldimann, 'The Iranian Nuclear Impasse', pp. 171–2; Ali Ansari, *Iran under Ahmadinejad: The Politics of Confrontation*, Adelphi Paper 393 (London: Routledge for the IISS, 2007).

73 Sayyid Ali Khamenei, 'Leader's Address to Executive Officials', The Center for Preserving and Publishing the Works of Grand Ayatollah Sayyid Ali Khamenei, 30 June 2007, http://english.khamenei.ir/index.php?option=com_content&task=view&id=198&Itemid=31.

74 *IRI Network*, 9 September 2008, BBC Monitoring, 10 September 2008. Shariatmadari has said that 'a brief study of the performance of the

reformists reveals that in many periods their positions were completely in line with the foreign enemy', *Keyhan*, 3 February 2008, BBC Monitoring, 4 February 2008; and *Keyhan* website, 13 January 2008, BBC Monitoring, 18 January 2008.

75 See Marlo Sogham, 'Iran's President and the Battle for the Arab Street', Radio Free Europe, 2 October 2008, http://www.rferl.org/content/Iran_President_Battle_Arab/1293381.html; Dan Morrison, 'Persian Populist Wins Arab Embrace', *Christian Science Monitor*, 21 June 2006, http://www.csmonitor.com/2006/0621/p06s01-wome.html.

76 Economic Research and Policy Department, Central Bank of the Islamic Republic of Iran, 'Annual Review 1383 (2004/05)', September 2005, available at http://www.cbi.ir/default_en.aspx. Quoted in Elliot Hen-Tov, 'Understanding Iran's New Authoritarianism', *Washington Quarterly*, vol. 30, no. 1, Winter 2006–07, pp. 172, 178. See also Mohammad Reza Bahonar, *Aftab-e Yazd*, 27 October 2008, BBC Monitoring, 3 November 2008. By one estimate, Iran's oil income averaged $21bn per year between 1997–2004, totalling $173bn, but between 2005–08 the total reached $200bn. Associated Press, 'Iran Feels Economic Pain as Oil Prices Fall', *International Herald Tribune*, 31 October 2008, http://www.iht.com/articles/ap/2008/10/31/news/ML-Iran-Feeling-the-Pain.php. An Iranian source goes further, asserting that the government's oil income in the past three years has exceeded the 'entire country's oil revenue in the history of Iran's oil'. *E'temad-e Melli*, 13 October 2008, BBC Monitoring, 15 October 2008. The IMF's figures tell a similar story, with oil revenues of about $17bn in 1999/2000 increasing to an estimated $66bn in 2007/08 and government revenues from oil going from 10% to 21% of GDP. See *Islamic Republic of Iran: Selected Issues*, IMF Country Report no. 08/285 (Washington DC, August 2008), p. 40. Other sources include *Iran: Facts and Figures*, Annual Statistical Bulletin, OPEC, 2007.

77 The figure $54bn is from Ann Fifield, 'Oil Price Fall Puts Squeeze on Iran's Cash', *Financial Times*, 23 October 2008, p. 4.

78 Robin Hughes, 'Iran Replenishes Hezbollah's Arms Inventory', *Janes Defence Week*, 3 January 2007, p. 17.

79 Associated Press, 'Iran Feels Economic Pain as Oil Prices Fall'. Notes Graham Fuller, 'despite considerable financial resources of its own, the loss of Iranian funding would significantly constrain [Hizbullah's] range of activities, especially its anti-Israeli guerrilla campaigns, even if the loss would not bring it to its knees.' Graham Fuller, 'The Hizbollah–Iran Connection: Model for Sunni Resistance', *Washington Quarterly*, vol. 30, no. 1, Winter 2006–07, p. 143.

80 One source estimates that Iran contributed $150m to Hamas during the second half of 2008; see 'Iran Pledges to Continue Support for Hamas', *Asharq Al-Awsat*, 26 May 2008, http://aawsat.com/english/print.asp?artid=id12877. For a discussion of Iranian contributions to both Hizbullah and Iraq (amounting in the latter case to $10m in grants and

$2bn in soft loans) see Barbara Slavin 'Mullahs, Money and Militias: How Iran Exerts its Influence in the Middle East', *USIP Special Report 206*, June 2008, pp. 11, 15. One assumes that Iran is also actively funding at least some Afghan warlords, if only as an insurance policy (as it did the Northern Alliance earlier).

81 The oil price slide, according to OPEC, is a 'dramatic collapse – unprecedented in speed and magnitude'. Carola Hoyos and Javier Blas, 'OPEC Cut Fails to Stop Oil Price Slide', *Financial Times*, 25–6 October 2008, p. 4.

82 Nader Habibi, 'The Iranian Economy in the Shadow of Economic Sanctions', *Middle East Brief*, no. 31, Crown Center for Middle East Studies, October 2008, http://www.brandeis.edu/crown/publications/meb/MEB31.pdf.

83 For a discussion of Iranian economic mismanagement see Najmeh Bozorgmehr, 'Iran Banks Struggle with Credit Shortages', *Financial Times*, 5 November 2008, p. 6.

84 Associated Press, 'Iran Feels Economic Pain as Oil Prices Fall'.

85 See Gareth Smyth, 'Iran Worried by Oil Price Falls', *Emerging Markets*, 12 October 2008; Mehr News Agency, report published by *Hezbollah News* website, 28 October 2008, BBC Monitoring, 5 November 2008; John Leyne, 'Iran Economy Facing "A Perfect Storm"', BBC News, 24 October 2008, http://news.bbc.co.uk/2/hi/middle_east/7687107.stm; Hossein Aryan, 'Falling Oil Compounds Iranian President's Problems', *Global Security.org*, 29 October 2008, http://www.globalsecurity.org/wmd/library/news/iran/2008/iran-081029-rferl01.htm.

86 Hade Nile, 'Tehran Begins to Feel the Pain of the Finance Crisis', *Washington Times*, 5 November 2008; Bernard Guetta, 'Couteaux tirés a Teheran', *Le Temps*, 18 October 2008, p. 2.

87 'UAE Banks Stepping up Pressure on Iranian Firms', *Bnet*, October 2008, http://findarticles.com/p/articles/mi_kmafp/is_200810/ai_n30943305.

88 Robin Wright, 'Using Banks to Stay a Step Ahead of Tehran', *International Herald Tribune*, 1–2 November 2008, pp. 15–16; Orde Kittrie, 'How to Put the Squeeze on Iran', *Wall Street Journal*, 14–16 November 2008, p. 13. See also Roger Stern, 'We can Beat Iran – But Not by Fighting', *International Herald Tribune*, 10 January 2008, p. 6.

89 A recent report by the US National Intelligence Council suggests that a drop in oil prices to $55–60 would put 'significant pressure' on the regime to choose between subsidising populist economic programmes and 'sustaining and funding … intelligence and security operations and other programs designed to extend its regional power'. *Global Trends 2025: A Transformed World* (Washington DC: National Intelligence Council, November 2008), pp. 45–51.

90 See Karim Sadjadpour, 'How Relevant is the "Iranian Street?"', *Washington Quarterly*, vol. 30, no. 1, Winter 2006–07, pp. 157–62; Simon Romero, Michael Slackman and Clifford Levy, 'Rich Oil Countries Face a Tighter Future', *International Herald Tribune*, 21 October 2008, pp. 1/8; 'An Axis in Need of Oiling', *Economist*, 25 October 2008, pp. 63–4; 'The Party's

Over', *Economist*, 22 November 2008, p. 48. 'Ordinary Iranians might take issue with the proposition that Iran wants to be a hegemon on the cheap', says Kuwaiti analyst Samir al-Frey. 'They would differ on the price Iran is paying.' Samir al-Frey, Middle East Institute Conference, Washington DC, February 2008.

91 Anthony Cordesman, *Conventional Armed Forces in the Gulf: An Overview*, Working Draft (Washington DC: CSIS, 23 June 2008), http://www.csis.org/media/csis/pubs/080623_gulfmilbal.pdf.

92 With antiquated technology, diminished production and growing domestic consumption, some analysts predict that Iran may not have *any* oil available for export within five years. Roger Stern, 'The Iranian Petroleum Crisis and United States National Security', *Proceedings of the National Academy of Sciences of the USA* (*PNAS*), vol. 104, no. 102, January 2007, http://www.pnas.org/content/104/1/377.full.pdf+html. The irony of Tehran relying on Saudi Arabia is not lost on observers; see Jean-Michel Bezat, 'The Drop in Oil Weakens Several Producing States', *Le Monde*, 26–7 October 2008, p. 13.

93 Fouad Ajami uses this phrase in *The Foreigners' Gift*, p. 253, in reference to Iraq.

94 See Tariq Alhomayed, 'Iran and Israel: The Escalation Continues', *Asharq Al-Awsat*, 1 July 2008, http://www.aawsat.com/english/news.asp?section=2&id=13258. Mark Mazzetti quotes Iraqis trained in Iran: 'Iranians believe they are superior to everyone' in 'US Report Details Iranian Role in Iraq', *International Herald Tribune*, 20 October 2008, p. 4.

95 Vali Nasr's characterisation of Iran in 2004 still seems applicable: 'Iran is currently a tired dictatorship teetering on the verge of collapse'. Vali Nasr, 'Regional Implications of the Shia Revival in Iraq', *Washington Quarterly*, vol. 27, no. 3, Summer 2004, p. 20. For a persuasive analysis, see Geoffrey Kemp, 'Our Imaginary Enemy', *National Interest*, May–June 2008. Says Kemp, 'Having the capacity to cause damage may give Iran a strong hand to play in regional geopolitics but this does not mean it can be a regional hegemon', pp. 33–4.

Is Pandemic Flu a Security Threat?

Christian Enemark

Most infectious diseases do not attract heightened political attention because their effects are mild, they are familiar to physicians, or their geographic occurrence is limited. A particular disease might be deemed a security issue, however, when its effects impose or threaten to impose an intolerable burden on society. That burden can be measured in terms of morbidity and mortality, but also in terms of the way in which a disease is perceived by those who fear infection. The disease described by the World Health Organisation (WHO) as 'the most feared security threat' today is pandemic influenza.[1] The next pandemic could cause illness and death on a large scale, over a wide area, in a short space of time. Such a prospect arguably sets this disease apart from the many others that may be regarded simply as health issues, and some Western governments have started to frame pandemic influenza as a threat to national security. According to the US pandemic plan, a 'necessary enabler of pandemic preparedness' is that this be viewed 'as a national security issue'.[2] The 'National Security Strategy of the United Kingdom' assesses an influenza pandemic as the 'highest risk' civil emergency.[3] And under the Australian pandemic plan, which emphasises 'maintenance of social functioning',[4] Australians are to receive the best possible health care 'commensurate with the maintenance of a safe and secure society'.[5]

Historical experience indicates that the world is overdue for an influenza pandemic, and a virus with pandemic potential – H_5N_1 avian influenza,

Christian Enemark is Lecturer at the Centre for International Security Studies, University of Sydney, and a Visiting Fellow at the John Curtin School of Medical Research, Australian National University.

Survival | vol. 51 no. 1 | February–March 2009 | pp. 191–214 DOI 10.1080/00396330902749798

which emerged in late 2003 – is still out of control. Past pandemics were all the more damaging because they took the world by surprise; however, the stakes are still high in the twenty-first century because increased human interconnectedness facilitates the global spread of disease. In the first identifiable pre-pandemic phase of human history, it makes sense to be thinking seriously about how best to prepare and respond. Yet as some governments move to prioritise this health issue by framing it in security terms, two risks emerge: firstly, that emergency responses implemented at the domestic level might do more harm than good; and secondly, that placing too great an emphasis on the health and security interests of individual states might detract from the need for long-term international cooperation on resisting pandemic influenza.

A pandemic in waiting?

The avian influenza virus known as H5N1 first appeared in Hong Kong chicken farms in 1997 and killed six out of 18 people infected when the virus jumped species. Although the properties of the virus were not well known at the time, the killing of all poultry in Hong Kong's markets and farms was a precaution that may well have averted a larger human outbreak of the disease. Thereafter, H5N1 was largely forgotten, but did not disappear. On 12 December 2003 South Korea's chief veterinary officer advised the World Organisation for Animal Health that a large number of chickens on a farm near Seoul had suddenly died of avian influenza. By early January 2004, reports were emerging of a 'mysterious disease' that had killed thousands of chickens in southern Vietnam.[6] In the months that followed, the influenza virus (by then identified as H5N1) swept through Asia, forcing authorities to cull tens of millions of domestic poultry. After the virus infected wild, migratory birds at Qinghai Lake in central China in mid 2005, it rapidly spread westward into Europe, India, the Middle East and Africa. By the end of 2008, H5N1 outbreaks in birds had occurred in 48 countries.[7]

The significance to human health of this avian-influenza strain lies in its potential to mutate into a form capable of sustained person-to-person transmission. H5N1 has repeatedly managed to infect humans, and every instance of this is a potential mutation opportunity. By late 2008 there had been 387 confirmed cases of human H5N1 infection across 15 countries since late 2003, including 245 deaths, for a global average case-fatality rate of around 63%.[8] To date, disease in humans has been caused by close contact with infected birds. In May 2006, however, a cluster of six deaths from H5N1 infection within one Indonesian family prompted an urgent WHO investigation, and a subsequent statistical study indicated that human-to-human disease transmission had probably occurred.[9] The world is presently at Phase 3 (of six phases) of the WHO pandemic-alert scheme: there are human infections with a new influenza subtype, but no confirmed human-to-human spread. The overarching public-health goals at this point are to ensure rapid characterisation of the new virus, and early detection and notification of and response to additional cases of human infection. Phase 6 would be the influenza pandemic itself: increased and sustained human-to-human transmission in the general population. The overarching public-health goal at that stage would be to minimise the impact of the pandemic.[10]

The worst pandemic of the twentieth century was the 1918 'Spanish Flu', which the WHO estimates killed 40–50 million people worldwide.[11] Subsequent pandemics in 1957 and 1968 were much less deadly, causing two million and one million deaths respectively.[12] The conservative estimate of the WHO, using epidemiological modelling based on 1957 data, is that a future influenza pandemic would cause 2–7.4m deaths worldwide.[13] However, another estimate based on 1918–20 data has predicted 62m deaths.[14] Concern about the severity of the next pandemic deepened after evidence emerged in 2005 that the Spanish Flu was caused by an avian influenza virus (H1N1) which, like H5N1, adapted to infect humans.[15]

The US government's 'Community Strategy for Pandemic Influenza Mitigation' designates five categories of pandemic influenza. All categories assume an illness rate among the total US population (around 300m) of 20–40%, with the mildest (Category 1) compared to seasonal influenza. This category has a case-fatality rate of less than 0.1%. A Category 2 pandemic would resemble the 1957 and 1968 influenza pandemics with a case-fatality rate of 0.1–0.5%. The worst-case scenario, a Category 5 pandemic comparable to the 1918 Spanish Flu, would have a case-fatality rate of 2%. Projected numbers of US deaths (less than 90,000 for Category 1 and over 1.8m for

Category 5) are based on the assumption of an unmitigated pandemic without intervention.[16] Based on this scheme, all things being equal, the predicted likelihood of any individual dying of an influenza infection during a pandemic would be 0.02–0.8%. This low risk of dying would presumably be reduced further by effective disease-control interventions.

The above figures indicate that an influenza virus in pandemic form would not generate case-fatality rates anywhere near as high as have been seen with bird-to-human H5N1 infections. Beyond mortality predictions, however, the health damage from pandemic influenza would be manifested more broadly in widespread non-fatal illness and general concern about infection. The WHO anticipates illness affecting around 25% of the world's population, or more than 1.5 billion people.[17] The essence of the global public-health challenge posed by pandemic influenza would thus be scarcity of resources. As the pandemic caused waves of outbreaks in humans lasting 1–2 months in a given region, completing its global spread in 8–12 months or less,[18] there would be a large number of sick people over a large geographical area, all requiring care during a short time period. The human and material resources for health care, which are usually stretched even in ordinary times, would be rapidly overwhelmed. In the event of a pandemic, the vast majority of those who fell ill would be able to recover with little or no medical assistance, but many others, seriously affected, would require prolonged hospitalisation, extensive drug treatment and sustained attention from well-trained medical professionals to survive.[19]

An influenza pandemic would cause immense economic disruption

Alongside high levels of illness and death, an influenza pandemic would cause immense economic disruption. An Australian study has estimated that a mild pandemic would result in 1.4m deaths and cost the global economy $330bn in lost economic output. The worst-case scenario was 142m dead and a loss to the global economy of $4.4 trillion.[20] Still, as grave as the health and economic consequences of a pandemic could be, there is reason to consider whether this threat can and should be framed in the politically supercharged language of 'security'.

Pandemic influenza and national security

The idea of linking health and security concerns as a matter of academic analysis and public policy has received support from two directions. In one camp, members of the public-health sector have adopted the language of security as a means of rallying political support and financial resources to address neglected health issues. For example, since the publication of the 1994 UN Human Development report, the notion of 'human security' has included 'health security'.[21] And the World Health Report of 2007 advocated the pursuit of 'global public health security', defined as 'the activities required, both proactive and reactive, to minimize vulnerability to acute public health events that endanger the collective health of populations living across geographical regions and international boundaries'.[22] Meanwhile, there are those within the security sector who argue that the impact of particular health challenges is serious enough to warrant the prioritisation traditionally accorded to the use of armed force. Infectious diseases have long been highly relevant to military operations. In April 2003, for example, when Canada's health minister suggested that Canadian Forces medical staff could help relieve pressure on Toronto hospitals then treating patients with severe acute respiratory syndrome (SARS), the military replied that the removal of such vital personnel would delay a major deployment to Afghanistan.[23] Another traditional dimension of military concern with infectious diseases is outbreaks resulting from human agency, and there is an abundance of literature on state and non-state interest in biological weapons.[24]

The most conspicuous issue in contemporary academic and policy literature linking infectious diseases to security is HIV/AIDS. The passage in 2000 of UN Security Council Resolution 1308 was the first time a health issue had been officially framed as an international security concern. The resolution expressed particular concern about the potential adverse effects of HIV/AIDS on UN peacekeeping missions, but it also stressed more generally that this disease, 'if unchecked, may pose a risk to stability and security'.[25] Analysis of HIV/AIDS by international-relations scholars typically spans a range of issues, from how the disease is undermining military capacity to how it is impoverishing millions and destroying social structures vital to internal state security.[26] Such arguments have been embraced in government circles as well. For example, HIV/AIDS featured prominently in an influential US National Intelligence Council report on the implications for the United States of global infectious-disease threats.[27] Other, more recent academic analysis of the links between health and security includes the work of Mely Caballero-Anthony on SARS, and Sara Davies and Alexander Kelle on international health governance.[28] With regard to pandemic influenza specifically, however, there have been few thoroughgoing attempts at analysing this disease in security terms.[29]

A case for framing pandemic influenza as a security issue can be made with reference to the potential societal consequences and likely perceptions of this threat to human health. In addition to the effects of human and economic losses, the two key elements in the security equation are speed and dread. With respect to the former, the widespread damage caused by a pandemic would seem all the worse because it happened so quickly. On this point, it is worth considering why it is that military threats are traditionally accorded the highest priority among national concerns. For Barry Buzan, the answer lies in the swiftness with which the use of armed force can inflict major undesired changes. 'Military action', he writes, 'can wreck the work of centuries in all other sectors. Difficult accomplishments in politics, art, industry, culture and all human activities can be undone by the use of force.'[30] More generally, for Richard Ullman, the essence of a security threat is 'an action or sequence of events that ... threatens drastically and over a relatively brief span of time to degrade the quality of life for the inhabitants of a state'.[31] Ole Waever defines a security problem as something that can, in a particularly rapid or dramatic fashion, undercut the political order within a state, deprive it of the capacity to manage by itself, and thereby 'alter the premises for all other questions'.[32] Just as states fear military conflict because so many national achievements could be quickly undone, so too an influenza pandemic would set back hard-won economic gains and potentially undermine trust in government. And like the all-consuming effort of prosecuting a war, defeating the flu would

Defeating the flu would become a first-order issue for governments

become a first-order issue for governments. This argument is reflected, for example, in the US pandemic plan which includes the statement: 'In terms of its scope, the impact of a severe pandemic may be more comparable to that of war or a widespread economic crisis than a hurricane, earthquake, or act of terrorism'.[33]

Related to the speed of an influenza pandemic is the dread it would instil as it raised both the individual fear of infection and the collective fear of contagion. Such dread could compromise the day-to-day human interactions which sustain modern societies: civilised humankind functions and survives largely because people have contact with and depend upon one another. The US government has warned that

> absenteeism across multiple sectors related to personal illness, illness in family members, fear of contagion, or public health measures to limit contact with others could threaten the functioning of critical infrastructure, the movement of goods and services, and operation of institutions such as schools and universities. A pandemic would thus have significant implications for the economy, national security, and the basic functioning of society.[34]

A pandemic-influenza virus would produce symptoms unfamiliar to most people, and the anxiety this generated would likely be compounded by the inability of medical professionals to provide adequate treatment. By inspiring dread, pandemic influenza is arguably likely to generate a level of societal disruption disproportionate to the health burden it poses. All things being equal, any given individual is extremely unlikely to succumb to pandemic influenza, yet the gap between real and perceived risk is likely to remain as a foundation for framing the threat as a security issue. Paul Slovic and his co-authors, writing in 1980 about risk perceptions, observed that '[society] appears to react more strongly to infrequent large losses of life than to frequent small losses'.[35] They identified 'dread' as a 'higher-order' characteristic of risk which, they argued, correlated closely with a strong societal desire for risk reduction.[36] The 2002 World Health Report referred to these findings and acknowledged, 'the higher the dread factor levels and the higher the perceived unknown risks, the more people want action to reduce these risks, including through stricter government regulation and legislative controls'.[37] Fear is disproportionately evoked by certain characteristics of threats, including involuntary exposure, unfamiliarity, and invisibility,[38] and a pandemic-influenza virus would exhibit all of these.

Historical experience suggests strongly that the speed and dread of pandemic influenza would indeed endanger internal state security. When a fast-spreading and unfamiliar disease inspires fear within national populations, the social contract under which citizens rely on governments to protect them during times of crisis is subjected to severe pressure. The SARS outbreak of 2003 provided a glimpse of this phenomenon when, in parts of China, there were riots caused by rumours of government plans to establish local SARS-patient isolation wards.[39] In 1994 an epidemic of plague in the Indian city of Surat engendered such terror that a quarter of the population fled within four days. This exodus, as Peter Chalk observed, 'fuelled an unprecedented level of anxiety across India, with fear and ignorance combining to freeze out even basic inter-personal sentiments of caring and civility ... So great was this national hysteria that the Delhi government was forced to bring in a police Rapid Reaction Force to effectively quarantine Surat.'[40] Such incidents demonstrate the panic caused when populations imagine a disease out of control, and where governments are seemingly incapable of securing the safety of their citizens.

Although it is impossible to predict precisely how the public would behave in response to the next influenza pandemic, it is clear that some governments expect severe social disruption as national health systems come under unprecedented pressure. In the United Kingdom, for example, contingency plans for a pandemic include posting police at doctors' surgeries and health clinics to stop panicking crowds from stealing medication.[41] And in the United States, the Department of Defense (DoD) plan for pandemic influenza includes the provision: 'When directed by the President, DoD will provide support to civil authorities in the event of a civil disturbance'.[42] It is anticipated that 'breakdowns in public order' might occur

> as health care facilities are overwhelmed with those seeking care and treatment for themselves or family members; as persons vie for limited doses of vaccines and antiviral medications; as supply-chain disruptions cause shortages in basic necessities; as individuals attempt to leave areas where outbreaks have occurred or where containment measures are in place.[43]

Such extraordinary use of military force in a domestic context is but one reflection of how a government might perceive pandemic influenza as having national-security significance, although such action also carries the risk of generating unintended adverse consequences. It is conceivable, for example, that the sight of troops on the street might exacerbate rather than assuage popular anxiety. More generally, the risk of approaching pandemic influenza as a security issue is that this could lead to emergency responses which are ineffective, counterproductive or unjust.

Risks of the national-security approach

As infectious diseases have made their way gradually onto the national-security agendas of governments in such places as the United States, the United Kingdom and Australia, some scholars have warned against the downsides of such a move. Susan Peterson, for example, has argued that public-health advocates might generate unwanted effects by appealing to the 'high politics' of security because this implies that 'human health is less important than, and can be justified only in terms of its impact on, security'.[44] Rebecca Katz and Daniel Singer observe that, while health professionals may welcome the higher profile and greater resources, approaching a health issue in security terms 'may … put relatively greater emphasis on the views of those outside the health community and potentially alter the approach to solving the problem'.[45] And Colin McInnes and Kelley Lee find it ironic that some public-health officials have emphasised the security implications of infectious diseases as a means of pushing health higher on policy agendas because this has resulted in 'increased concern, not for shifting patterns of health and disease of world populations as a whole, but for selected infections that potentially threaten the privileged few'.[46]

In a seminal paper published in 2006, Stefan Elbe considered the ethical implications of 'securitising' a specific disease: HIV/AIDS.[47] Elbe acknowledged that linking HIV/AIDS to security 'could accrue vital economic, social, and political benefits for millions of affected people by raising awareness of the pandemic's debilitating global consequences and by bolstering resources for international AIDS initiatives'.[48] An example might be the US President's Emergency Plan for AIDS Relief (PEPFAR). The original plan (announced in 2003) allocated $15bn over five years for HIV/AIDS treatment, prevention and care, and the recent 5-year PEPFAR renewal brings an additional $48bn.[49] At the same time, Elbe warned of the risk that 'framing the issue as a security issue pushes responses to the disease away from civil society toward the much less transparent workings of military and intelligence organizations, which also possess the power to override human rights and civil liberties – including those of persons living with HIV/AIDS'.[50]

Elbe's observations on the difficulty of reconciling health and security interests are relevant to pandemic influenza. Elevating pandemic influenza to the security agenda may serve to attract priority political attention, a higher level of resource allocation for public health, and the implementation of emergency-response measures. Given the scale and nature of the threat in question, these benefits should not be lightly dismissed. Nevertheless, it is incumbent on those within and outside government who would frame a pandemic in security terms to acknowledge the risks that accompany such a move.

Domestic responses

From a technical standpoint, pandemic influenza is fundamentally an issue of human health, and employing the tools of medicine and public health would be the most useful mode of response. The methods and institutions of 'security' are not necessarily compatible with these tools. Katz and Singer have observed that 'characterizing a health issue as a security threat often results in it being addressed through programmes and policies developed for law enforcement rather than public health'.[51] This could, for example, distort public-health ethics. Many interventions intended to stop the spread of disease – compulsory isolation, enforced quarantine, and restrictions on travel – could infringe individuals' civil and political rights. On the other hand, it could be sensible to impose some restrictions on the freedom of individuals in the interests of avoiding or mitigating what could be a disaster for many. Human rights are not absolute: international human-rights law has long recognised that 'public health may be invoked as a ground for limiting certain rights in order to allow a state to take measures dealing with a serious threat to the health of the population'.[52]

The legitimacy of restrictions invoked for public-health purposes, which may also be framed in security terms, need to be measured according to the latest and best-available scientific findings. However, historical experience indicates that populations can react to a disease outbreak without regard to science, and governments can respond in a manner contrary to public-health principles. During India's experience with bubonic plague from 1896 to 1914, the intrusion of British colonial authorities into the domain of private life was both aggressive and unprecedented. It involved entering homes, meddling with caste and religious practices, and regulating disposal of the dead. The effect of such intervention was often to intensify and quicken the panic occasioned by the disease itself. This in turn provoked fierce resistance, riots, mob attacks on Europeans, and even the assassination of British officials.[53] Admittedly, these efforts to formulate an effective plague policy were hindered by general ignorance about the causes and transmission modes of the disease as well as possible methods of treatment. Today, by contrast, a lot more is known about the microorganisms that cause infectious diseases, so popular fears may be lessened by the knowledge that effective medical treatment is available. Nevertheless, the emergence of diseases hitherto unknown to science has seen people and governments reverting to reactions based more on fear than evidence. Regarding state responses to HIV/AIDS, Elbe has observed periods of great insensitivity involving people living with HIV being 'ostracized and even persecuted by some states for their illness'.[54] He argues that framing the disease as a national-security threat 'risks fuelling such exclusionary and dehumanizing responses and could

The effect of intervention was often to intensify the panic

serve as an implicit legitimization of any harsh or unjust "emergency" policies that states may adopt in relation to persons living with the virus'.[55] Elbe's observations on HIV/AIDS are a warning that security-oriented, emergency measures to address infectious-disease threats must not in themselves curtail human rights to the point that such measures become illegitimate and counterproductive.

However, specific policy lessons learned from the experience of responding to HIV/AIDS are not readily applicable to pandemic influenza. This is because the two diseases have very different effects. HIV is hard to transmit and incubates within the body for a long period. Moreover, there is no vaccine available. By contrast, a pandemic-influenza virus transmits easily, has a very short incubation period, and can be protected against through vaccination. These differences generate unique response challenges. Responses to an influenza pandemic, once underway, generally fall into two categories: pharmaceutical (vaccines and antiviral drugs) and non-pharmaceutical (infection control and social distancing). The implementation of these responses may involve suspending or curtailing individuals' human rights, and the scarcity of resources is likely to require the prioritisation of medical benefits for some people over others. Arguably, the overall aim of promoting the greater good of society by way of public-health intervention justifies such violations and inequities when disaster would otherwise result. The important task is then to limit appropriately the occasions on which seemingly draconian public-health measures are invoked in the name of security.

Governments preparing for an influenza pandemic tend to focus on non-pharmaceutical responses as the most immediate. This is unsurprising as antiviral drugs to treat influenza and prevent its spread are in short supply, and a vaccine would not be available for several months after a pandemic virus started to spread. Non-pharmaceutical interventions for pandemic mitigation recommended by the US government include isolation in the home or a healthcare setting of all persons with confirmed or probable pandemic influenza; voluntary home quarantine of members of households with confirmed or probable influenza cases; dismissal of students from school and closure of childcare programmes; and the use of social-distancing measures to reduce contact between adults in the community and workplace, including cancellation of large public gatherings.[56]

Although the value of non-pharmaceutical responses to pandemic influenza is uncertain, governments may seek to maximise such responses in circumstances where nothing else is available to them. As Kelley Lee and David Fidler warn, 'public officials may feel compelled to adopt measures to demonstrate to domestic constituencies that they are "doing something".'[57] This was certainly the case in 2003 when governments confronted the new threat of SARS by implementing strict societal controls. As a consequence of the fact that no specific vaccine or treatment for SARS existed, the only interventions that seemed effective were those that minimised human contact. Transmission of SARS was quickly discovered to be containable using basic measures such as isolation of all infected patients and quarantine of close contacts, but government action also extended to the closure of inter-state borders and the use of military personnel to assist in enforcing containment orders.[58] In China, the legal and policy measures implemented to combat SARS were among the strictest in the world. Anyone who knowingly spread the disease could have faced capital punishment, and those who broke quarantine or evaded compulsory medical examination or treatment, and accidentally passed on the illness, faced up to seven years' imprisonment.[59] In Singapore the government instituted prison terms and large fines for anyone who broke quarantine.[60]

There is a risk that governments approaching pandemic influenza as a security issue may look to the SARS experience as a precedent for imposing similarly strict controls. It is important to note, however, that non-pharmaceutical measures such as quarantine and isolation are unlikely to be as effective against influenza as, in hindsight, they were against SARS. This is largely because the microorganisms that cause the diseases behave differently: influenza is far more contagious than SARS, has a shorter incubation period, and can be transmitted before the onset of symptoms.[61] Advice commissioned by the WHO on responding to pandemic influenza goes so far as to state that 'forced isolation and quarantine are ineffective and impractical'.[62] Furthermore, it is a common misconception that individual states can turn themselves into fortresses against pandemic influenza by closing their borders.[63] Indeed, border protection is emblematic of a narrow, national-security mindset being applied to a microbial threat that is inherently transnational in scope. Travel restrictions are unlikely to have much effect on the spread of a pandemic virus and are probably not practicable in most countries. During previous influenza pandemics, the screening and quarantining of travellers at international borders did not substantially delay the arrival of the pandemic, and similar policies today, 'even if they could be implemented in time and regardless of expense, would doubtfully be more effective in the modern era of extensive international air travel'.[64] On this point, it is worth noting that 35m people were screened for SARS at airports in four Asian countries in 2003, and not one case of the disease was detected.[65] Mathematical modellers have predicted that the pursuit of 'border security' through travel restrictions and health screening would bring negligible benefits.[66] Such an overly state-centric, security-oriented approach would also be expensive to implement and likely to generate immense opportunity costs in terms of reduced trade and travel.

If extraordinary efforts and additional resources are to be directed to responding to the threat of pandemic influenza, it makes more sense for governments to place greater emphasis on pharmaceutical responses as these are more promising for disease-control purposes. Of immediate use are antiviral drugs which, if taken shortly before or after a person is exposed to an influenza virus, can prevent that person from becoming infected and spreading the disease to others. A vaccine matched to the pandemic-influenza virus would take several months to produce once the pandemic commenced, but thereafter would be the single most effective defence. The trouble with pharmaceutical approaches, however, is that individual states cannot deploy them at will within their territory in the way they can non-pharmaceutical defences. Rather, states, as well as non-state entities, must compete in a global market for antivirals and vaccines. Under these circumstances, even the wealthiest states are unlikely to acquire as much medicine as they require as soon as they require it. As for the developing world, where 96% of all pandemic-influenza deaths are predicted to occur,[67] pharmaceutical defences are for the most part unaffordable or inaccessible.

In circumstances where non-pharmaceutical measures implemented domestically are of limited utility, and where acquiring national stockpiles of pharmaceuticals is difficult, a focus on national interest alone is inadequate. Against the worldwide threat of pandemic influenza, public health and security need also to be pursued cooperatively. The risk of placing too great an emphasis on the interests of individual states is that this might detract from the international openness and cooperation necessary to resist pandemic influenza most effectively in the long term.

International cooperation

The problem of reconciling traditional, state-based notions of security with challenges that are transnational in nature arose during the 1990s debates over 'environmental security'. Among those most stridently opposed to broadening the security agenda to include environmental degradation was Daniel Deudney. He argued, firstly, that pursuing national security and solving environmental problems are processes so fundamentally dissimilar that they cannot possibly fall under the single heading 'environmental security'.[68] Secondly, he insisted that employing the language of security to address environmental challenges could be counterproductive: 'For environmentalists to dress their programmes in the blood-soaked garments of the war system betrays their core values and creates confusion about the real tasks at hand.'[69] A decade later, addressing the issue of infectious diseases and referring to Deudney's warnings, Susan Peterson argued that linking an issue to security may raise awareness, 'but it likely also will hinder much of the cooperation that … public health advocates seek and that the disastrous humanitarian and development effects of [infectious diseases] demand.'[70] Similarly, for Harley Feldbaum and his co-authors, one of the reasons why the humanitarian objectives of the health field do not fit readily into the perspective of national security is that 'global health works for the benefit of all people within and across states, while national security prioritises the needs of one state over the needs of others'.[71]

In the World Health Report of 2007 the WHO declared that 'a truly effective international preparedness and response coordination mechanism cannot be managed nationally'.[72] This resonates with a basic principle contained in the 1946 WHO Constitution: 'The health of all peoples is fundamental to the attainment of peace and security and is dependent upon the fullest co-operation of individuals and States'.[73] Under the newly promulgated International Health Regulations (2005), states must notify the WHO of cases of 'human influenza caused by a new subtype' because these are 'events that may constitute a public health emergency of *international* concern' (emphasis added).[74] However, recent experience suggests that the pursuit of national self-interest can sometimes hinder the international cooperation necessary to address a transnational health threat. Calculated government opacity with the potential to jeopardise human health beyond national borders was a problem during the SARS outbreak of 2003. When people in southern China started succumbing to a mystery illness at the end of 2002, the initial response of the Chinese government was to deny the existence of a problem and to refuse offers of assistance from international health authorities. After 15 March 2003, when the WHO issued its first global warning about the SARS virus, China's government-controlled media was prohibited from reporting on the warning.[75] Not until 2 April 2003 was the WHO allowed to visit Guangdong Province to confirm that the alert it had received, nearly two months before, was consistent with cases of an entirely new disease by then appearing in hospitals around the world. When this became known, China was roundly criticised for not cooperating earlier with the WHO, as this failure to act may have allowed SARS to spread further. By the time the outbreak was declared over in July 2003, the disease had spread to 30 countries and killed 774 out of 8098 people infected.[76]

The pursuit of national self-interest can hinder international cooperation

China's initial response to SARS demonstrated how the withholding of information only compounds the damage from a fast-spreading infectious disease, hinders the task of containing an outbreak, and heightens the risk posed to other countries. However, Indonesia appeared not to have heeded this lesson when, in January 2007, it started withholding samples of H5N1 influenza virus from the WHO Global Influenza Surveillance Network (GISN). This was a major blow because virus samples must be supplied in a timely fashion for the GISN to serve its purpose of alerting countries to the emergence of influenza strains with pandemic potential. Indonesia, as the country most heavily afflicted by human H5N1 infections, is the most important source of information on the virus's properties and behaviour. The Indonesian government claimed that the WHO was transferring viral samples to pharmaceutical companies to make influenza vaccines for which the Indonesian people would have to pay an unacceptably high price. According to Indonesian Health Minister Siti Supari, the current global vaccine-supply scheme, which delivers doses almost exclusively to the developed world, is 'more dangerous that the threat of an H5N1 pandemic itself'.[77] In an attempt to secure an affordable vaccine supply in the event of an influenza pandemic, Indonesia subsequently commenced negotiations to sell H5N1 virus samples exclusively to the US-based company Baxter International.[78] At a meeting with WHO officials in March 2007, the Indonesian health minister sought assurances that her country would get a vaccine if a pandemic occurred. Indonesia subsequently agreed to negotiate on resuming the supply of virus samples to the WHO, subject to the agency seeking Indonesia's authorisation before sharing these with other researchers.[79]

The refusals by China and Indonesia to cooperate with the WHO and its member states are a warning against placing too much emphasis on national self-interest when dealing with the threat of pandemic influenza. The risk of deeming this threat a national-security issue is that a country will adopt a fortress mentality. Such an approach is futile against a disease that is impervious to political barriers. Rather, if pandemic influenza is to be resisted effectively into the long term, states need to act in their collective interest. There are some promising indications that wealthier countries are directing attention and resources beyond their own borders to protect themselves and others against pandemic influenza. The goals of the International Partnership on Avian and Pandemic Influenza, announced by US President George W. Bush in September 2005, include elevating the avian-influenza issue on national agendas; coordinating efforts among donor and affected nations; mobilising and leveraging resources; increasing transparency in disease reporting and improving surveillance; and building local capacity to identify, contain and respond to an influenza pandemic.[80] At a conference in Beijing in January 2006, the world's major aid donors led by the United States and the European Union pledged $1.9bn towards preventing an influenza pandemic. About half of this sum was to be spent in Vietnam, Laos, Cambodia, Indonesia and Thailand on strengthening disease surveillance, improving laboratory and health services, and boosting the communication capacity of these countries.[81] An additional $406.1m was pledged at an international conference in New Delhi on 4–6 December 2007.[82]

Another example of extraordinary political commitment at the international level arose in response to Indonesia's drastic action on H5N1 virus samples. At the Sixtieth World Health Assembly in May 2007, the WHO member states passed a resolution requesting that the organisation 'establish an international stockpile of vaccines for H5N1 or other influenza viruses of pandemic potential, and to formulate mechanisms and guidelines aimed at ensuring fair and equitable distribution of pandemic influenza vaccines at affordable prices in the event of a pandemic'.[83] In combination with pledges of financial aid from the developed to the developing world, this gesture may indicate that states recognise the paramount importance of international cooperation against an emerging microbial threat that transcends political borders. However, it remains to be seen if the spirit of cooperation will withstand the ravages of a pandemic itself. If individual states do choose to see this as a national-security crisis, the emergency-response measures implemented should support rather than subvert the public-health interests of other states.

* * *

If H5N1 or some other influenza subtype mutates into a form transmissible between humans, the result could be illness and death on a large scale, over a wide area, in a short space of time. Fearing swift and severe economic disruption, a loss of public confidence, and the undermining of societal functioning, governments may choose to frame pandemic influenza as a first-order issue of national security, one which would alter the premise for all other activity. The United States, Britain and Australia are already moving in this direction. In adopting such an approach there is potential to rally political support for improving and mobilising public-health resources. However, there is also a risk that addressing a health issue in security terms will lead to emergency responses which are ineffective, counterproductive or unjust. Overzealous non-pharmaceutical responses to a pandemic, such as forced quarantine and border screening, would likely have little effect in controlling the disease yet could undermine human rights and exacerbate economic losses. In addition, it is important to look beyond the vulnerability of individual states and their purely domestic actions taken in response. Appealing to national security may be a good way of focusing national attention and resources. However, given the transnational reach of this potential microbial threat and the increasing interdependence of states, the greater part of the solution must lie in international cooperation. To resist pandemic influenza most effectively in the long term, all states have a collective interest in sharing health information, improving local public-health capacity, and expanding the global supply of influenza vaccine.

Notes

1 World Health Organisation, *A Safer Future: Global Public Health Security in the 21st Century, World Health Report 2007* (Geneva: World Health Organisation, 2007), p. 45.

2 *Implementation Plan for the National Strategy for Pandemic Influenza* (Washington DC: Homeland Security Council, 2006), p. 18.

3 'The National Security Strategy of the United Kingdom: Security in an Interdependent World', United Kingdom Cabinet Office, 19 March 2008, p. 14, http://interactive.cabinetoffice.gov.uk/documents/security/national_security_strategy.pdf.

4 Australian Department of Health and Ageing, *Australian Health Management Plan for Pandemic Influenza* (Canberra: Commonwealth of Australia, 2006), p. 19, http://www.health.gov.au/internet/main/publishing.nsf/Content/ohp-pandemic-ahmppi.htm/$file/AHMPPI.pdf.

5 *Ibid.*, p. 51.

6 'Mysterious Disease Kills Thousands of Vietnamese Chickens', *Sydney Morning Herald*, 7 January 2004, http://www.smh.com.au/articles/2004/01/06/1073268033540.html.

7 World Organisation for Animal Health, 'Update on Avian Influenza in Animals (Type H5)', 4 December 2008, http://www.oie.int/downld/AVIANINFLUENZA/A_AI-Asia.htm.

8 World Health Organisation, 'Confirmed Human Cases of Avian Influenza A (H5N1)', 10 September

2008; http://www.who.int/csr/disease/avian_influenza/country/en/.

9 Yang Yang, M. Elizabeth Halloran, Jonathan D. Sugimoto and Ira M. Longini, 'Detecting Human-to-Human Transmission of Avian Influenza A (H5N1)', *Emerging Infectious Diseases*, vol. 13, no. 9, September 2007, p. 1351.

10 World Health Organisation, *WHO Global Influenza Preparedness Plan: The Role of WHO and Recommendations for National Measures Before and During Pandemics* (Geneva: World Health Organisation, 2005), p. 2.

11 World Health Organisation, 'Ten Things You Need to Know about Pandemic Influenza', 14 October 2005, http://www.who.int/csr/disease/influenza/pandemic10things/en/.

12 *Ibid.*

13 *Ibid.*

14 Christopher J.L. Murray, Alan D. Lopez, Brian Chin, Dennis Feehan and Kenneth H. Hill, 'Estimation of Potential Global Pandemic Influenza Mortality on the Basis of Vital Registry Data from the 1918–20 Pandemic: A Quantitative Analysis', *The Lancet*, vol. 368, no. 9554, 23 December 2006, pp. 2211–18.

15 Jeffery K. Taubenberger, A.H. Reid, R.M. Lourens, R. Wang, G. Jin and T.G. Fanning, 'Characterization of the 1918 Influenza Virus Polymerase Genes', *Nature*, vol. 437, no. 7060, 6 October 2005, pp. 889–93. Notwithstanding these findings, it is important to acknowledge that mutation of H5N1 into a pandemic form is not certain to occur. At present, the virus attaches to cells in the lower respiratory tract of humans. It is possible that a feature of this particular virus may prevent it from ever attaching to cells in the upperrespiratory tract and thus transmitting easily to others through coughing and sneezing. See Debby van Riel, Vincent J. Munster, Emmie de Wit, Guus F. Rimmelzwaan, Ron A.M. Fouchier, Ab D.M.E. Osterhaus and Thijs Kuiken, 'H5N1 Virus Attachment to Lower Respiratory Tract', *Science*, vol. 312, no. 5772, 21 April 2006, p. 399.

16 US Department of Health and Human Services, 'Community Strategy for Pandemic Influenza Mitigation', February 2007, http://www.pandemicflu.gov/plan/community/commitigation.html.

17 World Health Organisation, *A Safer Future*, p. 16.

18 Klaus Stöhr and Marja Esveld, 'Will Vaccines Be Available for the Next Influenza Pandemic?', *Science*, vol. 306, no. 5705, 24 December 2004, p. 2195.

19 Jaro Kotalik, 'Preparing for an Influenza Pandemic: Ethical Issues', in Michael J. Selgelid, Margaret P. Battin and Charles B. Smith (eds), *Ethics and Infectious Disease* (Malden: Blackwell, 2006), p. 97.

20 Warwick McKibbin and Alexandra Sidorenko, *Global Macroeconomic Consequences of Pandemic Influenza* (Sydney: Lowy Institute for International Policy, 2006).

21 UNDP, *Human Development Report* (New York: United Nations Development Programme, 1994), pp. 27–8.

22 World Health Organisation, *A Safer Future*, p. 5.

23 Shannon Selin, 'The Security Implications of SARS', *CANCAPS Bulletin*, no. 37, 2003, p. 10.

24 Recent publications include Susan Wright, 'Terrorists and Biological Weapons: Forging the Linkage in the Clinton Administration', *Politics and the Life Sciences*, vol. 25, no. 1–2, March–September 2006, pp. 57–115; Reynolds M. Salerno and Lauren T. Hickok, 'Strengthening Bioterrorism Prevention: Global Biological Material Management', *Biosecurity and Bioterrorism*, vol. 5, no. 2, June 2007, pp. 107–16; Nicholas A. Sims, 'The Future of Biological Disarmament: New Hope after the Sixth Review Conference of the Biological Weapons Convention', *Nonproliferation Review*, vol. 14, no. 2, July 2007, pp. 351–72. The use of influenza virus for biological-warfare purposes, highly unlikely but not impossible, is beyond the scope of this article.

25 United Nations Security Council, UNSCR 1308, 17 July 2000, http://daccessdds.un.org/doc/UNDOC/GEN/N00/536/02/PDF/N0053602.pdf.

26 See, for example, P.W. Singer, 'AIDS and International Security', *Survival*, vol. 44, no. 1, Spring 2002, pp. 145–58; Gwyn Prins, 'AIDS and Global Security', *International Affairs*, vol. 80, no. 5, October 2004, pp. 931–52; Stefan Elbe, *Strategic Implications of HIV/AIDS*, Adelphi Paper 357 (New York: Oxford University Press for the International Institute for Strategic Studies, 2003); Alan Dupont, *East Asia Imperilled: Transnational Challenges to Security* (Cambridge: Cambridge University Press, 2001), pp. 212–27; Harley Feldbaum, Kelley Lee and Preeti Patel, 'The National Security Implications of HIV/AIDS', *PLoS Medicine*, vol. 3, no. 6, June 2006, pp. 0774–78; Colin McInnes, 'HIV/AIDS and Security', *International Affairs*, vol. 82, no. 2, March 2006, pp. 315–26.

27 National Intelligence Council, 'The Global Infectious Disease Threat and its Implications for the United States', http://www.wilsoncenter.org/topics/pubs/Report6-3.pdf, January 2000.

28 Mely Caballero-Anthony, 'Combating Infectious Diseases in East Asia: Securitization and Global Public Goods for Health', *Journal of International Affairs*, vol. 59, no. 2, Spring 2006, pp. 105–27; Alexander Kelle, 'Securitization of International Public Health: Implications for Global Health Governance and the Biological Weapons Prohibition Regime', *Global Governance*, vol. 13, no. 2, April–June 2007, pp. 217–35, Sara E. Davies, 'Securitizing Infectious Disease', *International Affairs*, vol. 84, no. 2, March 2008, pp. 295–313.

29 Julian Palmore, 'A Clear and Present Danger to International Security: Highly Pathogenic Avian Influenza', *Defense and Security Analysis*, vol. 22, no. 2, June 2006, pp. 111–21; Christian Enemark, *Disease and Security: Natural Plagues and Biological Weapons in East Asia* (London: Routledge, 2007), pp. 41–58.

30 Barry Buzan, *People, States and Fear: An Agenda for International Security Studies in the Post-Cold War Era*, 2nd ed. (London: Harvester Wheatsheaf, 1991), p. 117.

31 Richard Ullman, 'Redefining Security', *International Security*, vol. 8, no. 1, Summer 1983, p. 133.

32 Ole Waever, 'Securitization and Desecuritization', in Ronnie D. Lipschutz (ed.), *On Security* (New

York: Columbia University Press, 1995), p. 52.

33 *Implementation Plan for the National Strategy for Pandemic Influenza*, p. 2.

34 *Ibid.*, p. 1.

35 Paul Slovic, Baruch Fischhoff and Sarah Lichtenstein, 'Facts and Fears: Understanding Perceived Risk', in Richard C. Schwing and Walter A. Albers (eds), *Societal Risk Assessment: How Safe is Safe Enough?* (New York: Plenum, 1980), p. 209.

36 *Ibid.*, p. 211.

37 World Health Organisation, *World Health Report 2002: Reducing Risks, Promoting Healthy Life* (Geneva: World Health Organisation, 2002), p. 32.

38 Jessica Stern, 'Dreaded Risks and the Control of Biological Weapons', *International Security*, vol. 27, no. 3, Winter 2002–03, p. 102.

39 Erik Eckholm, 'SARS is the Spark for a Riot in China', *New York Times*, 29 April 2003, p. A1.

40 Peter Chalk, 'Disease and the Complex Processes of Securitization in the Asia-Pacific', in Mely Caballero-Anthony, Ralf Emmers and Amitav Acharya (eds), *Non-Traditional Security in Asia: Dilemmas in Securitization* (Aldershot: Ashgate, 2006), p. 127.

41 Stewart Tendler, 'Flu Doctors to be Given Police Guards', *Times Online*, 2 November 2005, http://www.timesonline.co.uk/article/0,,25149-1853843,00.html.

42 Department of Defense, *Department of Defense Implementation Plan for Pandemic Influenza* (Washington DC: Office of the Assistant Secretary of Defense for Homeland Defense, 2006), p. 16.

43 'Implementation Plan for the National Strategy for Pandemic Influenza', p. 153.

44 Susan Peterson, 'Epidemic Disease and National Security', *Security Studies*, vol. 12, no. 2, December 2002, p. 51.

45 Rebecca Katz and Daniel A. Singer, 'Health and Security in Foreign Policy', *Bulletin of the World Health Organization*, vol. 85, no. 3, March 2007, p. 233.

46 Colin McInnes and Kelley Lee, 'Health, Security and Foreign Policy', *Review of International Studies*, vol. 32, no. 1, January 2006, p. 11.

47 Stefan Elbe, 'Should HIV/AIDS Be Securitized? The Ethical Dilemmas of Linking HIV/AIDS and Security', *International Studies Quarterly*, vol. 50, no. 1, March 2006, pp. 119–44.

48 *Ibid.*, p. 120.

49 Nellie Bristol, 'US Senate Passes New PEPFAR Bill', *The Lancet*, vol. 372, no. 9635, 26 July 2008, p. 277.

50 Elbe, 'Should HIV/AIDS Be Securitized?', p. 128.

51 Katz and Singer, 'Health and Security in Foreign Policy', p. 234.

52 'Siracusa Principles on the Limitation and Derogation Provisions in the International Covenant on Civil and Political Rights', *Human Rights Quarterly*, vol. 7, no. 1, 1985, p. 6.

53 Rajnarayan Chandavarkar, 'Plague Panic and Epidemic Politics in India, 1896–1914', in Terence Ranger and Paul Slack (eds), *Epidemics and Ideas: Essays on the Historical Perception of Pestilence* (New York: Cambridge University Press, 1992), pp. 8–10, 204.

54 Elbe, 'Should HIV/AIDS Be Securitized?', p. 128.

55 *Ibid.*

56 US Department of Health and Human Services, 'Community Strategy for Pandemic Influenza Mitigation'.

57 K. Lee and D. Fidler, 'Avian and Pandemic Influenza: Progress and Problems with Global Health Governance', *Global Public Health*, vol. 2, no. 3, July 2007, p. 226.

58 Elizabeth M. Prescott, 'SARS: A Warning', *Survival*, vol. 45, no. 3, Autumn 2003, p. 218; *World Health Report 2003: Shaping the Future* (Geneva: World Health Organisation, 2003), p. 77.

59 Jacques deLisle, 'SARS, Greater China, and the Pathologies of Globalisation and Transition', *Orbis*, vol. 47, no. 4, Fall 2003, p. 598.

60 Ann Mongoven, 'The War on Disease and the War on Terror: A Dangerous Metaphorical Nexus?', *Cambridge Quarterly of Healthcare Ethics*, vol. 15, no. 4, October 2006, p. 412.

61 World Health Organisation Writing Group, 'Nonpharmaceutical Interventions for Pandemic Influenza, International Measures', *Emerging Infectious Diseases*, vol. 12, no. 1, January 2006, p. 83.

62 *Ibid.*, p. 88.

63 See, for example, Athol Yates, 'Close Airports or Pandemic Influenza Will Fly In', Australian Homeland Security Research Centre, August 2007, http://www.homelandsecurity.org.au/files/airport_closure_flu.pdf.

64 WHO Writing Group, 'Nonpharmaceutical Interventions for Pandemic Influenza, International Measures', p. 86.

65 'Comments from the Center for Biosecurity of UPMC on the National Strategy for Pandemic Influenza: Implementation Plan', *Biosecurity and Bioterrorism*, vol. 4, no. 3, 2006, p. 322.

66 Peter Caley, Niels G. Becker and David J. Philp, 'The Waiting Time for Inter-Country Spread of Pandemic Influenza', *PLoS ONE*, vol. 2, no. 1, art. e143, January 2007.

67 'International Health Regulations: the Challenges ahead', *The Lancet*, vol. 369, no. 9575, 26 May 2007, p. 1763.

68 Daniel Deudney, 'The Case against Linking Environmental Degradation and National Security', *Millennium*, vol. 19, no. 3, December 1990, pp. 462–65.

69 *Ibid.*, p. 475.

70 Peterson, 'Epidemic Disease and National Security', p. 46.

71 Harley Feldbaum, Preeti Patel, Egbert Sondorp and Kelley Lee, 'Global Health and National Security: The Need for Critical Engagement', *Medicine, Conflict and Survival*, vol. 22, no. 3, July–September 2006, p. 196.

72 World Health Organisation, *A Safer Future*, p. 66.

73 World Health Organisation, 'Constitution', 1946, http://www.who.int/entity/governance/eb/who_constitution_en.pdf.

74 World Health Organisation, '58th World Health Assembly, International Health Regulations (2005)', 2005, http://www.who.int/csr/ihr/WHA58-en.pdf.

75 Yanzhong Huang, 'Mortal Peril: Public Health in China and its Security Implications', *Health and Security Series* (Washington DC: Chemical and Biological Arms Control Institute, 2003), p. 69.

76 World Health Organisation, 'Summary of Probable SARS

Cases with Onset of Illness from 1 November 2002 to 31 July 2003', 26 September 2003, http://www.who.int/csr/sars/country/table2003_09_23/en/.

77 Dennis Normile, 'Indonesia to Share Flu Samples Under New Terms', *Science*, vol. 316, no. 5821, 6 April 2007, p. 37.

78 John Aglionby and Andrew Jack, 'Indonesia Blames WHO for Bird Flu Deal', *Business Day Online*, 8 February 2007, http://businessdayonline.com/?c=125&a=11127.

79 'Indonesia Edges Closer to Sharing Bird-Flu Samples', *Nature*, vol. 450, no. 7170, 26 November 2007, p. 599.

80 US Department of State, 'US Government Support to Combat Avian and Pandemic Influenza', 4 June 2007, http://www.state.gov/r/pa/scp/86190.htm.

81 Margaret Harris Cheng, 'Cash Boost for Avian Influenza Exceeds Expectations', *The Lancet*, vol. 367, no. 9507, 28 January 2006, p. 289.

82 'Chair's Summary', New Delhi International Ministerial Conference on Avian and Pandemic Influenza, December 2007, available at http://www.delmincon.com/.

83 World Health Organisation, 'World Health Assembly Closes', 23 May 2007, http://www.who.int/mediacentre/news/releases/2007/wha02/en/index.html. Implementing the WHO plan to increase the global supply of vaccine is estimated to cost $3–10bn: World Health Organisation, *Global Pandemic Influenza Action Plan to Increase Vaccine Supply* (Geneva: World Health Organisation, 2006).

Review Essay

History, Tragedy and Farce

Michael Howard

The White War: Life and Death on the Italian Front 1915–1919
Mark Thompson. London: Faber and Faber, 2008. £25.00.
464 pp.

In the spring of 1945 the Allied Armies in Italy at last levered the Wehrmacht out of the mountain south of Bologna which they had held all winter, swept across the Po valley and headed past Venice into the Julian Alps en route for Vienna. My own unit had a more limited role: to get to Trieste and the valley of the Isonzo before Tito's partisans did. We spent the next year sitting in the bleak hinterland of Trieste to prevent the Yugoslavs from 'liberating' the territories that the Italians had 'liberated' from the Austrians 30 years earlier. We had not yet heard of the term 'ethnic cleansing' but we saw what it meant, close up. The Italians had done their best to 'cleanse' the Slavic population from the land they regarded as *Italia irredenda*: now the Yugoslavs were repaying in kind, and we could do little to stop them.

This was my own introduction to the Brave New World we had been fighting to create. But we also found a different kind of testimony to the region's tormented past. On our way through Friuli we briefly halted in the village of Redipuglia outside the largest war memorial I have ever seen, equal to if not dwarfing those on the Western Front: a vast staircase of terraces, each a hundred yards wide, containing the bones of over a hundred

Michael Howard is President Emeritus of the International Institute for Strategic Studies.

Survival | vol. 51 no. 1 | February–March 2009 | pp. 215–220 DOI 10.1080/00396330902749806

thousand Italian soldiers who had died on those huge, and to us unknown, battlefields in the First World War.

All that I, like most of my countrymen, then knew about the Italian performance in that war was that their army had shamefully disintegrated at the battle of Caporetto in 1917 and had to be rescued by British and French troops; a performance they had repeated more recently when they had tamely surrendered to the forces of General Wavell in the Western Desert. We had no idea that between 1915 and 1917 the Italian army had launched 11 successive offensives across the stony plateau of the Carso that now confronted us; attacks on the same scale, and with much the same results, as our own Battle of the Somme. When finally in November 1917 an Austro-German army launched their counter-offensive the Italian army had already been eviscerated. By the end of the war it had lost nearly half a million dead, most of them on the Carso. In 1918 Italy was able with Allied help to defeat an Austrian army as exhausted as her own and earn her place at the top table at the peace conference. But although the Italian prime minister declared that Italy's victory was 'one of the greatest that history has ever recorded', nobody, least of all his allies, believed him, or treated the country as if it were. In fact Italy had been as broken by her 'victory' as Germany by her defeat.

Perhaps it has been a combination of native shame and foreign contempt for the disastrous performance of the Italian armies that accounts for the remarkable absence of any serious study in English of the Italian front in the historiography of the First World War; but now the gap has been filled by Mark Thompson with this brilliant and definitive book. It is a work not simply of military but of political and, even more importantly, cultural history as well. All these dimensions are needed if we are to understand why the diverse, heterogeneous and often barely literate peoples of the Italian peninsula came to be pitch-forked, by a deeply divided political class dominated by a small group of ultra-nationalist fanatics, into a war of choice against her former allies of the Triple Alliance; and even more remarkably, how the very disaster of the war would strengthen that group so effectively that they were able to seize power and hold it until they were overthrown in a yet more disastrous war 30 years later.

The political reasons for Italy's entry into the war are clear enough. In 1914 her political leaders still had unfinished business. Her wars of unification in 1859 and 1866 had still left extensive Italophone territory south of her 'natural frontier' the Alps under control of the Austrian Empire. The Western Allies were thus able to bribe her out of neutrality by offering her, at the expense of the Austrians, not only these *irredenta,* but non-Italian territory beyond the Adriatic that would make her a major Balkan power; together with a hefty slice of Anatolia at the expense of her old enemy Turkey. The mouth-watering prospect of such Great-Power status was too great for the Italian cabinet to resist, and the nationalist enthusiasm stoked up by a hysterical press too great for its opponents to overcome. The cry that had once inspired the followers of Garibaldi, *fuori i stranieri!,* drive out the foreigners!, carried all before it. Good liberals, not only in Italy but throughout the Western world, believed that Italian honour demanded no less.

The trouble was that such 'good liberals' were rather thicker on the ground abroad than they were in Italy itself. In fact the Italian people had never been very enthusiastic about unification. Nationalism was largely a middle-class phenomenon and in Italy the middle class was slow to develop. The bulk of the population were still very largely peasants – overwhelming so in the south – who took their cue from a Catholic Church who saw a unified secular Italian state as a challenge to the political and spiritual authority of the Pope. The conservative ruling classes in the north were quite content with the enlarged Kingdom of Savoy bequeathed to them by Cavour, and regarded everything south of Rome – as many of them understandably still do – as a liability rather than an asset. Mazzini, the great father of Italian nationalism who spent most of his life in exile, constantly lamented the lack of enthusiasm among the actual inhabitants of the Italian peninsula for his great ideal of 'Italy'; and his followers were only too conscious of the fact that when 'Italy' was in fact made and the foreigners driven out, it was as a result, not of any great popular uprising, but of deals between her more powerful neighbours after wars in which the Italians themselves had played a minor and largely inglorious part. The great cry of Italian patriots was *Italia fara da se* – Italy will do it on her own. But she never yet had.

In consequence Italian nationalists believed that the Italy in which they lived was not the true Italy to which they aspired. That had still to be made, and could only be made by war – not a war won by proxy, but one in which Italians would fight and die, preferably in large numbers; showing by their sacrifice the greatness of the nation that had borne such heroes. Of course the Italians were not the only people who felt like this in 1914. The same ideals drove thousands of young British, German and French patriots to the recruiting offices and ultimately to their deaths, and the numbers in which they died were seen as evidence, not of the incompetence of their generals, but of the virility of their nations. Nor were Italian intellectuals alone in their espousal of the creed of 'vitalism'; that denial of all the values of the Enlightenment, which taught that action was its own justification and that both intellectual reflection and traditional morality were matter only for wimps; a twisted kind of Social Darwinism that saw nature as a struggle for survival in which thousands had to die in order that the species best adapted for conflict could prevail. But nowhere else were such ideas driven to the verge of lunacy as they were in Italy by the Futurists under Filippo Marinetti; or well beyond it, as they were by that ludicrous but sinister figure Gabriele d'Annunzio, in whom romantic excess reached its furthermost bounds and who in any other country would probably have been quietly locked up, instead of becoming a national icon.

This lethal combination of frenetic nationalism and anti-rational vitalism, burning away at the core of an otherwise somnolent and inward-looking people, helps to explain not only why Italy joined in the war in 1915, but why she fought it in the way that she did. Her overt objectives were the Friuli plain northeast of Venice, the mountainous region of South Tyrol (including Alto Adige, most of which was German in language and culture) and the port of Trieste; Italian-speaking, but entirely surrounded by Slovenes. The first objective was overrun almost bloodlessly, but its inhabitants greeted their liberators with a marked lack of enthusiasm. The second involved fighting in the high Alps that showed the Italians at their heroic and ingenious best; a campaign fought largely in bitter cold and perpetual snow that provides the title for Mark Thompson's book. But that was as untypical an experience for the bulk of the Italian army as was the Palestine campaign for the British. In the same way as the typical experience of the war for the British was the Western Front, the campaign in which the Italians died in their scores of thousands was that fought on the stony plateau of the Carso that barred the way to Trieste.

The Carso is a horrible place; a bare, gnarled, stony plateau bereft of cover, baking in summer, freezing in winter, and tilted at an angle ideal for providing defensive positions with excellent fields of fire. The Italian mode of attack was, to put it mildly, unsubtle: repeated frontal attacks by infantry en masse. The Italian High Command shared the belief, common throughout the European military in 1914, that the advantage provided to the defensive by modern firepower could always be overcome by the high morale of the attacking infantry. But by 1915, when Italy entered the war, both their allies and their enemies were already learning the bleak lesson laconically expressed by General Philippe Pétain – *fire kills*; and were beginning to explore less lethal ways of conducting war. Their learning process was painfully slow, but in the Italian army there was no learning process at all, and their High Command saw no reason to acquire any. Thompson's account of these attacks is quite heart-rending. The officer corps consisted largely of Savoyard monarchists who had little interest in the welfare of the largely illiterate peasants from the south from which the army was recruited. (It is significant that when Italian infantry attacked, their battle cry was not *Italia!* but *Savoia!*) At their head was General Cadorna, who blamed the failures of his offensives on a lack of *slancio* ('dash') on the part of his troops, and sacked any of his subordinates who suggested otherwise. Again, the Italians were not unique in this. Left to themselves, both Joffre and Haig took the same line. But they were not left to themselves: political pressure ultimately forced Joffre to retire and Haig to accept more realistic advisers on to his staff. But the Italian government stood firmly behind Cadorna, his parliamentary critics were howled down by nationalist fanatics, and the press meekly submitted to being muzzled by the High Command. The war in fact saw the destruction of Italian parliamentary democracy: Mussolini would simply administer the *coup de grâce*.

If the war was a tragedy for Italy, the peace was an ugly farce. Her allies had now been joined by the United States, who did not recognise the validity of the promises made to her by France and Britain; certainly not of those involving non-Italian territory now claimed by Yugoslavia. The Italian government had kept quiet about these in its wartime propaganda. Now they declared that if they were not granted regions of Dalmatia and Slovenia in which the Italians were a tiny minority, the heroes whose bodies lay on the blood-soaked slopes of the Carso would have died in vain. Italy's allies treated these claims with contempt. The seizure of the port of Fiume, which did not even figure in Italian war claims, by a private army under Gabriele d'Annunzio was a farcical piece of operatic bravura that even the Italian government had to disown. But the government found itself increasingly helpless in face of the *fasci* (bands) of hardline demobilised troops, termed by their leader Mussolini a 'trenchocracy' who claimed to represent the true Italy, with a mission to cleanse the political Augean stables at home and assert Italian greatness abroad.

For the British the lesson taught by the Somme and Passchendaele was 'never again'. For the mass of Italians the comparable holocaust on the Carso was a meaningless disaster better forgotten, but the Fascists claimed that it proved Italy to be truly a warrior nation, one that would now in future *fara da se*. But for one man at least, lessons had been learned. General Giulio Douhet had been one of the few open critics of Cadorna's conduct of the war and was determined that nothing of the kind should occur again. In 1921 he published a seminal book *The Command of the Air* which was to transform strategic thought. In future the target for military offensives should be not the armies of the enemy, but his civil population. There would be no more Carsos: war would in future be a matter for heroic elites, warriors destroying the soft centres of enemy power from afar. It was a doctrine eagerly embraced by Mussolini. Others were not slow to follow.

We were billeted in a smelly little village on the edge of the Carso when we learned over the radio of the bombing of Hiroshima. Then I saw no connection. I do now.

Book Reviews

Middle East

Steven Simon

Negotiating Arab–Israeli Peace: American Leadership in the Middle East
Daniel C. Kurtzer and Scott B. Lasensky. Washington DC: United States Institute of Peace Press, 2008. £9.95/$16.50. 210 pp.

Negotiating Arab–Israeli Peace serves two useful purposes. It sheds additional light on the murky world of the 'Middle East Peace Process', as it was pursued fitfully by the Clinton and Bush administrations, while also serving as a negotiating primer for the Obama administration. The two authors situate the failures of the past two administrations to mediate a final-status agreement between Israelis and Palestinians within the broader flow of US activity in the region, which, in their view, produced an 'alarming pattern of mismanaged diplomacy'. They argue that alternating bouts of US inattention and clumsy intervention have 'weakened the U.S. position in the region and on the world stage', even as it has damaged prospects for an eventual peace agreement.

Kurtzer and Lasensky return to the Camp David talks of 2000, which preceded the al-Aqsa intifada that erupted in the autumn of that year. Their assessment tracks with that of the first of the peace-team renegades, Robert Malley, who asserted publicly in 2001 that the US negotiators had been hasty, unprepared and insufficiently attentive to Yasser Arafat's political constraints, while preoccupied with Ehud Barak's domestic political requirements. The hothouse atmosphere in which peace-process negotiating strategies were cultivated, the authors argue, made informed decisions less likely. The small group of policymakers, led by Dennis Ross, was separate from the bureaucracy that handled day-to-day US Middle East policy. This was 'too insular and inhibited the development of

 DOI 10.1080/00396330902749814

US positions on core issues'. To make matters worse, the peace-process cadre 'scrambled at the last minute' to formulate negotiating positions on hellishly complicated issues like the future of Jerusalem. In the authors' view, the final nail in the coffin of Camp David II was hammered in by the Clinton administration's decision to place the failure entirely on Arafat's weakening shoulders. The Bush administration proved unwilling to re-engage before 11 September 2001 because of domestic political disincentives and the conviction that it was just a 'shepherd's war' with little consequence for US strategic interests. After 11 September, the 'war on terrorism' shaped foreign policy. On one hand, that was something that the United States and Israel could agree upon. On the other, the need to win Muslim support necessitated President Bush's June 2002 endorsement of a Palestinian state.

In essence, Kurtzer and Lasensky maintain that a sharp 16-year tilt in Israel's direction, during which, according to another renegade peace-process official, the US functioned as 'Israel's lawyer', had seriously damaged the peace process. In the authors' view, negotiations produce durable agreements when the parties see themselves as more or less equal. Titanic power imbalances, like that which exists between Israel and the Palestinians, tend to undercut the weaker party's incentive to cut a deal. The nature of US support for Israel's position has intensified the imbalance and weakened Palestinian interest in negotiating a deal that they fear will both reflect and perpetuate their inferior strategic position. The only way to change this dynamic is for Washington to embrace a somewhat more even-handed approach to mediation. With the Israeli Defense Force at the gates of Gaza City as the United States undergoes a presidential transition, the issues raised by this book will be central to any effort to reduce the overall level of violence and renew negotiations.

Negotiating for the Past: Archaeology, Nationalism, and Diplomacy in the Middle East, 1919–1941
James F. Goode. Austin, TX: University of Texas Press, 2007.
£12.99/$22.95. 307 pp.

Archaeology in the Middle East has long been used by both nationalists and colonialists to validate their respective claims. Napoleon's army in Egypt campaigned with archaeologists in tow, as did the French and British armies in the Levant in the next century. Neo-Orientalists who cherish their David Roberts lithographs of romantic Middle Eastern ruins should thank the British expeditionary force that brought the famous surveyor to Baalbek and other exotic locales. By laying claim to the physical remnants of ancient indigenous cultures, the conquerors were subordinating them as precursors to their own superior civilisation, while asserting control over the territory from which these remnants were extracted. Not surprisingly, the rise of nationalism in the region carried within it the impulse to reclaim these legacies. And just as the disparity between the glory of the ruins and wretchedness of the contemporary landscape confirmed the colonialist posture of superiority, the magnificence of the ruins was interpreted by nationalists as the harbinger of a glorious future. These battles are far from over and are not confined to the Middle East, as the see-saw battle between the United Kingdom and Greece over the Elgin Marbles attests. In 1996, the opening of the so-called Hasmonean Tunnel under the Temple Mount/Haram al-Sharif spurred riots that claimed dozens of lives. For Israelis, the tunnel was important because it reinforced a historical Jewish claim to the spot. For Arabs who reject this claim and even challenge the veracity of Biblical and other non-scriptural sources that support it, the excavation of the tunnel was an affront. Even more offensive was the presumed 'fact' that the Israeli dig would undermine the structural integrity of the Muslim holy places that sit astride the hill. Israelis have relied heavily on archaeology to buttress their right of return and counter ideologically driven assertions that the presence of Israelites or Jews in Palestine has never been more than fleeting, if not invented entirely. It was also useful in inculcating immigrants from many different countries with a sense of shared history and destiny. Hence the assiduous excavation and documentation of tells throughout Israel, the West Bank and Gaza and the astoundingly thorough excavation of the walls and environs of the Temple ruins.

James Goode's book explores the struggle between colonialists and nationalists for ownership of the past in Turkey, Egypt, Iran and Iraq. Using an impressive array of archival evidence, drawn from government and university records, as well as memoirs and biography, Goode brings the commitments of archaeologists on both sides of the struggle to life. The case he makes is that in the inter-war period the control of archaeological sites and the artefacts taken from them gradually shifted from foreign to local control. Yet in some ways, the transition was more de jure than de facto, since the position of antiquities director in key countries, such as Egypt and Iran, remained in the hands of Western archaeologists, while the training of local archaeologists was generally not encouraged by the Western institutions that were in effective control of antiquities departments. The situation was rather more complicated than even this suggests, because some of the American and European scholars involved were champions of local control over archaeological sites and their riches, even as their colleagues expressed themselves in what we would consider to be racist terms. Goode's history dexterously sorts through these complexities, while establishing a palpable sense of the times and places in which events occurred.

Foxbats over Dimona: The Soviets' Nuclear Gamble in the Six-Day War
Isabella Ginor and Gideon Remez. New Haven, CT: Yale University Press, 2008. £10.99/$17.00. 304 pp.

The conventional wisdom about the origins of the 1967 Arab–Israeli war held that the Soviets mistakenly informed the Egyptians that Israel had massed a dozen or so Israeli Defense Force divisions along the Syrian border in preparation for a surprise attack. This inaccurate warning then triggered the positioning of Arab armies to pre-empt the expected Israeli assault. As the crisis unfolded, the Soviets retracted their intelligence reporting on an impending Israeli attack and worked feverishly to convince all parties to stand down. These efforts proved fruitless in part because of Egyptian confidence in their ability to deal Israel a serious a defeat and Israeli threat perceptions, fuelled by the fear that Egyptian forces might reach and destroy the nuclear-weapons facility at Dimona. This concern was spurred by what many had thought were overflights of Dimona just before the war by Egyptian air force MiG-21s. Or so it was thought until Ginor and Remez came out with this book. It turns out, as some had long suspected, that the conventional wisdom seems mostly to have been a function of Soviet disinformation.

Based on newly accessible Soviet archives, and, more importantly, scores of interviews with former Soviet officials and military personnel who were involved in these dramatic events, the authors argue that the 1967 war occurred as a result of Soviet machinations, rather than as an unintended consequence of bad intelligence, badly timed. According to Ginor and Remez, the Soviets deliberately fed their Egyptian client false information about an Israeli build-up to spur Egypt's mobilisation, which the Soviets assumed would trigger an Israeli pre-emptive attack that would ineluctably result in an Arab victory.

By now, dear reader, you are probably asking yourself, 'What could the Soviets possibly have been thinking?' Ginor and Remez answer that question by unpacking the principles of Soviet strategy in the region. From the Soviets' perspective, the key to their influence with the Arab states lay in the ability to deter Israeli aggression through the implicit threat of a Soviet nuclear response. It followed that the minute Israel developed its own counter-deterrent, a Soviet response to an Israeli attack against Moscow's Arab clients would no longer be credible. Under these circumstances, Arab capitals would cease to rely on their Soviet patron, with predictably adverse implications for Soviet clout in the Middle East. This was an outcome that – from Moscow's standpoint – had to be prevented. Hence the elaborate plot to start a war intended to result in the elimination of Israel's nuclear installation at Dimona. The Soviets, as the title of this book indicates, did more than just deploy false intelligence to stir things up. The aircraft that overflew Dimona before the war were not, as had been thought, Egyptian MiG-21s; they were *Foxbats*, a much more advanced jet, piloted by Soviet crews. In the event, Israel's stunningly swift destruction of Egypt's ground and air forces rendered the Soviets' plan irrelevant. This is a genuinely interesting book that will lead many to revise their understanding of the 1967 war.

Inside Egypt: The Land of the Pharaohs on the Brink of a Revolution
John R. Bradley. Abingdon: Routledge, 2008. £14.99/$24.95.
242 pp.

Middle East watchers are blessed by the vigour of journalistic reporting about the region. Christopher deBellaigue in Iran, Scott Mcleod in Egypt, and Max Rodenbeck on transnational developments all offer readers based outside the region a sense of the gritty realities of life and politics on their respective beats. John Bradley, editor of *Arab News* and a fixture on the Cairene scene, is yet another spirited, intelligent voice from the Middle East. Having written a somewhat gloomy but absorbing book about Saudi Arabia, he has now turned his attention to Egypt. This is a deeply troubled country, according to Bradley. A labour-supply overhang, owing to a burgeoning population of young people entering the job market and an economy that is only just keeping pace, presents real dangers. The government lacks legitimacy, and its inability to deliver better living standards for the vast bulk of the population and unwillingness to open up political space, coupled with a propensity for human-rights violations, is fuelling widespread resentment. To keep a lid on this cauldron, the government employs upwards of 1.4 million security officers, while periodically arresting hundreds of opposition politicians and other activists. Thus far, this has proved a workable strategy.

Bradley's confidence in the durability of this approach, however, is low. Applying a kind of wave theory to the history of Egyptian politics, he points to three twentieth-century upheavals: in 1919, when there was a popular uprising against the British, who effectively ruled the country; in 1952, when the free officers overthrew the monarchy and established a republic; and in 1977, when mass protest again price increases led to Anwar Sadat's 'humiliating' reversal. Bradley sees momentum building toward another tumultuous turn. The fact that Egypt's president, Hosni Mubarak, is 80 years old and that there is no settled succession plan makes for a more brittle situation, one that is all the more susceptible to shock. Bradley acknowledges the front-runner status of Mubarak's allegedly reform-minded son, Gamal, but underscores his unpopularity in the street as well as the military and economic elite's scorn for economic liberalisation of the kind that Gamal is said to endorse. The implication is that Gamal, were he to succeed his father, would enjoy no greater legitimacy and would be hamstrung by the 'deep state'. Against the continued immiseration of ordinary Egyptians, this set of conditions could lead to the fourth great upheaval in modern Egyptian history. And coming roughly 30 years after the last one, it would fit the pattern quite neatly. This time, however, the Muslim Brotherhood, the only well-organised opposition movement with deep ties to the street and 88 members in the Egyptian parliament, will be waiting in the wings. Not that the Brothers are universally loved. Their attempts to suppress Mawlid festivities and grass-roots Sufism has not endeared them to many Egyptians. But they have the networks, funds and penetration of society on their side. For Bradley, the feared outcome is, in effect, a clerical revolution that turns Egypt into something resembling Iran.

Bradley's wave theory, as well as his sense of the fragility of Egyptian politics, will leave many unconvinced. Thus far, the state has demonstrated the will and capacity to retain power. Indeed, one of the underappreciated malign effects of the Iraq War was the strengthening of authoritarianism in Egypt. But one does not have to buy Bradley's conclusions to learn much from his intimate feel for Egypt, its people and politics.

The Yacoubian Building
Alaa al Aswany. Humphrey Davies, trans. New York: Harper Perennial, 2007. £7.99/$13.95. 272 pp.

Beaufort
Ron Leshem. Evan Fallenberg, trans. London: Harvill Secker, 2008. £12.99/$24.00. 311 pp.

The Yacoubian building is a Cairo landmark that Alaa al Aswany, a novelist and part-time dentist, uses as the setting for his complex novel about contemporary life in Egypt. Read alongside Bradley's book, it deepens one's understanding of the harshness of urban life in Egypt, while reminding readers that life goes on nevertheless. The characters represent a cross-section of society: the ageing manipulative businessman and his scheming, frustrated sister; the grasping and unscrupulous gangster and aspiring politician; the poor shopgirl, exploited at work and at home; a pair of artful-dodger brothers, trying to enlarge their miniscule shanty; and the disappointed Islamist. This dramatis personae implies that the book is a mere Dickensian comedy of manners. It is more than this. The ruthlessness with which the weak are abused by those in positions of relative power, as well as the determined resistance of those at the bottom and their determination to rip off the system, or their better-off neighbours, evokes state–society dynamics in Egypt today. The book works as metaphor on many levels. The fact that much of the action takes place in a mini-barrio on the rooftop of the Yacoubian building suggests that the degraded and corrupt conditions al Aswany describes have proliferated atop the ruins of pre-revolutionary Egypt. Of particular interest to *Survival* readers is al Aswany's depiction of the transformation of a marginal young man, trying to improve his lot by qualifying as a police recruit, who is thwarted because he does not have the money to bribe his examiners. Stripped of options, he gravitates to a radical mosque and, ultimately, to terrorism. His trajectory, as portrayed by al Aswany, vivifies the socio-economic roots of radicalism.

Beaufort is an Israeli novel and winner of the prestigious Sapir Prize. The author is a media figure with a keen ear for language. The book, which is about an Israeli unit preparing for the evacuation of an Israeli Defense Force installation atop a mediaeval fortress in southern Lebanon, is especially relevant now. In the first instance, it's a story about young men in combat faced with an implacable foe, in the form of Hizbullah, and exposed to serious risk in defending a fortress that they know will shortly be abandoned in the context of the Israeli withdrawal from Lebanon in May 2000. The narrative is studded with army slang, rich and pungent. But it is also a book about the changing nature of Israeli patriotism and, more broadly, society. An optimist is dismissed as a 'Herzl', while meddlers are scorned as 'Zionists'. Military duties are approached with grim determination and mistrust of the politicians who send troops into battle. Yet despite the bleakness of soldiers' attitudes, morale somehow persists, along with a powerful commitment to the job and, evidently, the state. This is a richly textured novel that illuminates the complexities of Israel today. As Israelis continue to rely heavily on their army, *Beaufort* should be required reading for Middle East watchers interested in the performance of these troops in a changing strategic setting.

Politics and International Relations

Gilles Andréani

The United Nations Security Council and War: The Evolution of Thought and Practice since 1945
Vaughan Lowe et al., eds. Oxford: Oxford University Press, 2008. £90.00/$200.00. 793 pp.

At no time in its history has the UN Security Council's role been more intensely debated than it has since the mid 1990s. The council's failure to act in time to stop the genocide in Rwanda and the mass killings in Srebrenica, its mixed record in implementing the disarmament it had mandated in Iraq after the 1991 Gulf War, its divisions over the second Iraq War in 2003, and its inability to agree on the reform of its membership have given rise to accusations of all sorts from a variety of opposed quarters. The American right, which had fiercely attacked the United Nations in the 1990s, proclaimed it dead after the 2003 crisis; the liberal interventionists devised doctrines to circumvent it and to use force without its consent for humanitarian purposes; and emerging countries decried the lack of legitimacy stemming from its outdated membership and the veto privilege enjoyed by its permanent members.

Against this background, a wider perspective on the Security Council is needed, one which can provide historical depth, data and analysis to the debate. The purpose of *The United Nations Security Council and War* is to provide just such a perspective. As an exhaustive and informed record of the council's activity in the field of international peace and security, from its creation in 1945 to the present, the book is a magnificent achievement. Edited by several eminent UN specialists, it includes more than 20 contributions, mostly from academics (alongside a few practitioners), grouped under four main headings. The first part is devoted to the historical and legal framework of the Security Council. The second covers the council's role, as it has evolved over time, often in ways unforeseen by the authors of the UN charter. This part addresses such themes as peacekeeping, sanctions, the council's relations with regional security arrangements, and its post-Cold War role. The third part provides analyses of a dozen case studies (Korea, Suez, the Arab–Israeli wars, India–Pakistan, East Timor, Iran–Iraq, the two Iraq wars, the former Yugoslavia, Bosnia, Afghanistan and West Africa) complemented by a chapter on instances of non-action by the council. The fourth and last part is entitled 'The Security Council and the Changing Character of War'; it addresses the role of the council with respect to humanitarian law, humanitarian intervention, the administration of territories, terrorism and the private use of force. Useful and well-documented appendices on, among other things, UN peacekeeping operations, UN-authorised operations and sanctions, vetoed resolutions, and armed conflicts since 1945 complement the main text.

The result of an ambitious and much-needed effort, this book will stand out as an indispensable tool in the vast literature on the United Nations Security Council, set apart by the quality of its research, the wealth of extensive and carefully researched data it contains, as well as the diversity of viewpoints it offers. The resulting picture one gets of the UN Security Council is rich, diverse, extremely stimulating and almost complete – one may regret the absence of a chapter on arms control. The authors are modest or wise enough to leave us without a conclusion on a subject both controversial and in flux. But their joint introduction, in itself remarkably like a small book on the Security Council, will invite all those with an interest in the subject to stay clear from simplistic anti-UN opinions and to conclude, as the editors put it at the end of their introduction, that 'the council remains as it was in 1945, undependable, unaccountable and unrepresentative [but also] a place of hope, a place to do some serious business, and a place few states would do without, reform or no reform'.

The UN Security Council and the Politics of International Authority
Bruce Cronin and Ian Hurd, eds. Abingdon: Routledge, 2008.
£22.99. 249 pp.

This volume, which belongs to a series devoted to security and governance, conveys much the same general message as *The United Nations Security Council and War*, that is, that the UN Security Council is more powerful and efficient than is generally acknowledged. It proceeds, however, from a totally different perspective, that of international-relations theory. Rather than starting from an exposition of cases and from the evolution of the council's practice over its history in order to draw broader lessons, it starts from one of the most difficult and elusive concepts of political science, that of authority, to make the case that the council has effectively developed its own authority in a number of fields, and somehow managed to constrain the autonomy of states.

The UN Security Council and the Politics of International Authority starts with an introductory analysis of international authority, defined as 'a social relation where a hierarchical relationship in the international sphere is recognized as legitimate', and its relevance to the Security Council's role according to previous international-relations research. Next, contributors explore the possible sources of the council's authority: either it emerges from a process of delegation by states, or is a function of the council's ability to reflect the evolving legal consensus among the members of the internal community. Two chapters are then devoted to important recent developments in the evolution of the council's authority: its role as a legislature, as evidenced by resolutions 1373 and 1540, which followed the 11 September 2001 attacks, and its normative role, which, according to contributor George J. Andreopoulos, has placed the council in danger of overstretch as a result of the expanded definition of what constitutes a threat to peace and security and hence a legal basis for the council to take action. In Part III, entitled 'The Exercise of council Authority', the book explores two new ways in which the council's role has manifested itself: in the creation of international criminal tribunals, and in relations with NGOs, who stood to gain much from the development of council-mandated peace operations. A final chapter explores how the council's authority has been circumvented, either through the 'united for peace' resolution or otherwise.

Books on international-relations theory are notoriously difficult to read, and research in this field tends to become ever more specialised, alienating the general readership. *The UN Security Council and the Politics of International Authority*, while making abundant reference to previous studies, contains interesting and original arguments on an important subject, and is commendably easier to read and less arcane than most international-relations publications.

The Thin Blue Line: How Humanitarianism Went to War
Conor Foley. London: Verso, 2008. £14.99/$26.95. 272 pp.

In an age of local conflicts, failed states and festering crises, the description of evolving situations on the ground and factual details tend to be more interesting than general theories. This is especially true of the question of humanitarian intervention. Endless debates pit those who approve of it based on the universal character of human values and rights against those who oppose it as an infringement on the sovereignty of states and the independence of peoples, or else as a new form of Western imperial dominance. This debate is equally divisive among governments, lawyers, scholars and humanitarians themselves: two former leaders of Doctors Without Borders, for instance, have staked out opposing positions on the question. The debate, however, is much less likely to be settled in theory than in practice. What good does humanitarian activism actually do? Does it effectively alleviate the tragedies of contemporary conflicts and demonstrably bring help to innocent civilians? Does it promote a measure of justice and peace?

The Thin Blue Line, by Conor Foley, a humanitarian worker with a lifetime of field experience in Kosovo, Iraq, Sudan, Indonesia, Afghanistan and Sri Lanka, mixes consideration of local situations with a general view that humanitarian intervention has crossed the 'thin line' that separates the disinterested provision of help to victims from the forceful imposition of Western solutions on unwilling peoples.

Foley's depiction of the shortcomings of the various operations in which he participated is honest and vivid. His discussions of the violence inflicted on the Serb and Roma minorities by Albanian-speaking extremists in Kosovo after the NATO air campaign, the fast dissipation of the goodwill of the Afghan population towards international aid workers as the counter-insurgency campaign developed, the manipulation of emergency relief by the Mengistu government in Ethiopia, and the mixed, if not adverse, impact of International Criminal Court (ICC) indictments against Lord's Resistance Army activists on the peace negotiations in Northern Uganda, all effectively illustrate the dilemmas and shortcomings of humanitarian intervention. By contrast, Foley sees the good record of Aceh after the peace agreement as only marginally related to the presence of international observers or humanitarians, whose daily life there he describes as mostly dull and of little use.

Altogether, the book is easy to read and insightful, describing many instances when humanitarian activism found its limits. At times it is even funny, as when Foley quotes a defence of humanitarian intervention by the UK Foreign Office in the mid 1990s: 'the best case that can be made in support of humanitarian intervention is that it cannot be said to be unambiguously illegal'. Foley raises many useful questions, in particular on the accountability of humanitarian activities, to which a rich chapter is devoted. Does this material vindicate Foley's overall very pessimistic assessment of humanitarianism? Not quite. In a conclusion entitled 'Towards a New Imperialism?' – an excessive title even with the question mark – he no longer seem so sure of himself and of what needs to be done. He calls for more modesty from the West and a recognition of others' values. He also calls for more, rather than less, intervention in the form of a more independent ICC, which includes the crime of aggression in its statute. But the limited character of these answers does not diminish the value of this book as a testimony of the intractable dilemmas raised by humanitarian action.

Moral Limit and Possibility in World Politics
Richard M. Price, ed. Cambridge: Cambridge University Press, 2008. £17.99/$32.99. 309 pp.

Over the years, constructivism has become an important tool for analysing international relations. Looking at the relations between states from a sociological standpoint, constructivism holds that beliefs and norms, as social constructs, play an important role in shaping international relations over time, alongside the quest for power and the desire for security, the two main immutable driving forces according to the realist school. The two schools do overlap to some extent, and both have had difficulties integrating moral judgement into their approach. The realists do not deny that morality has a part to play, but it is a qualified morality, that of politics, which must first and foremost account for the consequences of decisions: following the classical Weberian distinction, realists insist that it is the morals of responsibility that matter among states rather than those of conviction. (Raymond Aron stretches the argument in *Peace and War* by likening ethics among states to those which exist among gangsters.)

If one believes, however, that moral judgements are the product of time- and culture-bound constructions, as constructivists do, one finds oneself on even thinner ground than the realists when it comes to accounting for a moral dimension, and for the possibility of a certain moral progress, in international relations. The purpose of *Moral Limit and Possibility in World Politics* is to bridge that gap, and to offer a constructivist view of the possibility of moral change in world politics.

It does so convincingly. After three solid (and slightly scholarly for the non-specialist) introductory chapters, in which the problem is framed in theoretical terms, the book offers a collection of excellent essays on the moral dilemmas inherent in such issues as the recognition of indigenous peoples, US policy towards undocumented migrants, sanctions on Iraq, humanitarian interventions, gender equality and multiculturalism, and inter-state community building.

Taken together, these essays constitute a remarkably convergent set of views, displaying broad agreement that there is something akin to moral progress at work in world politics. Of course, one may call it progress only inasmuch as hypocrisy, the tribute paid by vice to virtue, can be taken as a form of progress. As the social recognition of higher norms of conduct, morality is indeed capable of influencing the beliefs and behaviours of actors, including in international politics.

In the latter portion of the book, the editor takes us from the exploration of how moral norms arise to the question of 'what to do' when facing the ethical dilemmas outlined in the previous chapters. There, the relativist bias inherent in the self-definition of constructivism looks like an insuperable obstacle. But it also reminds us that in a global and diverse world moral compromises are fragile, and ethical solutions are bound to be transitory in nature. A constructivist contribution to international ethics should not differ widely from the realist concern that politics should choose not between good and evil, but 'between the preferable and the detestable', to again quote Raymond Aron.

Global Ethics: Anarchy, Freedom and International Relations

Mervyn Frost. Abingdon: Routledge, 2008. £18.99/$35.95.
182 pp.

The current focus on morality in international debates is not necessarily an unmitigated good. There is a danger that moral arguments will be biased or misused as they were during the debate on the Iraq War, and that cynicism will set in as a result. Too much ethical talk, especially when divorced from the conduct of policy, can induce many to discard ethical considerations as a valid dimension in international relations. What we need in this context is not so much moral clarity as intellectual clarity about morality in international affairs.

This is what Mervyn Frost, professor of international relations and head of the Department of War Studies at King's College, offers us in his concise, well-written book on the current role of ethics in international relations. *Global Ethics* is a book for the general reader, praiseworthy for staying clear from international-relations jargon and from the internal epistemological controversies that plague the discipline. The book is nonetheless ambitious, since Frost offers a cohesive and forceful view that ethical dilemmas are actually central to the conduct of international politics, and makes a distinct contribution to better identifying, theorising and hopefully solving these dilemmas.

Frost starts with a description of the international system as an anarchical society of states, not just because this is a basic tenet of the English school of international relations, but because this model preserves the freedom and diversity of its constituent members. Alongside the international system, he identifies a global civil society as well, itself an anarchical society. Rather than opposing the global society of states with global civil society, he judges that institutional compromises which integrate both in international decision-making, on the EU model, is the best way to structure international relations. The rights of the individual as citizen under the state system, and as member of the global civil society, have to be combined, and opposing one to the other is a false moral dilemma.

Within this general framework, Frost addresses instances of ethical dilemmas, ranging from counter-terrorism to the use of private military companies, and to the decision to invade Iraq and then to stay there. On Iraq, he addresses the various sides of the moral debate, and uses the dual tenets of diversity and autonomy to conclude on the moral impropriety of the decision to go to war in Iraq, as well as of attitudes which tend to lay the blame for the current situation at the feet of the Iraqi government.

This short description of some of the basic propositions of this important book do not do justice to the trenchant, creative and nuanced thinking which is reflected in *Global Ethics*. The book will stand out as an excellent contribution to the essential study of ethics in contemporary international politics.

Arms, Arms Control and Technology

Bruno Tertrais

A Nuclear Family Vacation: Travels in the World of Atomic Weaponry
Nathan Hodge and Sharon Weinberger. London: Bloomsbury, 2008. £12.99/$24.99. 324 pp.

Survival readers may be surprised that a book called *A Nuclear Family Vacation* found its way into these pages. But as the saying goes, one should not judge a book by its cover – especially one which features, in the UK edition, a man lazily lying on a beach, wearing a swimming suit and gas mask, and watching a nuclear mushroom rising in the distance – as this travelogue is one of the most insightful recent texts on nuclear weapons.

The book recounts the voyages of Sharon Weinberger, author of *Imaginary Weapons: A Journey Through the Pentagon's Scientific Underworld*, and her husband Nathan Hodge to some of the most secretive places on Earth. The authors visited Russia, Kazakhstan and Iran, but most of their travel was within the United States, to parts of the country's nuclear complex: military commands, secret bunkers, weapons factories and test sites. The book's best moments include a visit to the US Strategic Command, which vividly reveals the marginalisation of nuclear options in American strategy and where officers are keener to quote Thomas Friedman than, say, Thomas Schelling or Lawrence Freedman; an encounter with one of the last survivors of *Operation Paperclip* (which saw German scientists flown by the United States to Huntsville, Alabama, in 1945); and accounts of repeated but failed attempts to visit Russian nuclear installations, where secrecy and paranoia still reign, perhaps in part to mask the extent of the corruption that pervades the channelling of international assistance to weapons dismantlement.

The book offers a rare blend of personal adventures, often narrated with dark humour, and insightful analysis of the challenges of maintaining an enormous industrial, technological and military base, with the associated know-how, in an era when the threat of immediate massive destruction by an enemy state has all but disappeared, whatever one thinks of the current debate about a 'resurgent' Russia. The authors are well-informed journalists but readily admit to having no particular knowledge about nuclear policy; this makes their conclusion all the more interesting. Their thesis is that the nuclear complex is driving on autopilot, without a clear sense of purpose, or rather, 'many purposes, but no particular end' (p. 284). The complex is also without leadership – it is true that the White House has shown surprisingly little interest in nuclear weapons during the past four years. The authors correctly note that 'there is no single person – or agency – who can put it all together' (p. 282). At a time when some researchers claim that the United States, for the first time, has an unchallenged nuclear superiority, the nuclear enterprise that Hodge and Weinberger describe is one that is in fact decaying. They found 'a complex adrift, grasping for meaning and purpose' (p. 277), 'an enterprise which had lost its way' (p. 281).

It may be hard to maintain what the Bush administration called a 'responsive infrastructure' without programmes such as the Reliable Replacement Warhead that Congress so far has been reluctant to fund. Will President Barack Obama change this state of affairs? It will be a challenging task. He has promised to put the goal of nuclear abolition, dear to liberals and neo-Reaganites alike, back on the agenda. Managing to revitalise the US nuclear enterprise while simultaneously promoting the eventual elimination of nuclear weapons would be a real tour de force.

The Long Shadow: Nuclear Weapons and Security in 21st Century Asia
Muthiah Alagappa, ed. Stanford, CA: Stanford University Press, 2008. $29.95. 517 pp.

Since the 2006 North Korean nuclear test, Asia is the continent that hosts the highest number of nuclear-capable countries – at least four (six if one includes Russia and the United States) out of a total of nine. It is therefore not surprising that many commentators, including this reviewer, have suggested for a long time that the nuclear future will be written in Asia.

Alagappa's ambition is to consolidate in a single volume of nearly 600 pages all the issues and problems related to the possession of nuclear weapons in Asia. It is a first, and his effort must be highly commended. *The Long Shadow* is an unprecedented collection of essays covering all the main countries and topics related to nuclear policies in Asia. Many country chapters elegantly summarise national visions of nuclear weapons. China's policy is described as 'Dynamic Minimum Deterrence', while India's is 'Assured Destruction', and North Korea's 'Existential Deterrence'. Others cover subjects that are rarely treated in international-strategic literature, such as Australia's current nuclear policy and vision of nuclear weapons. Particularly interesting is the chapter on the risks of nuclear terrorism, which analyses the possible motivations of major groups from the Abu Sayyaf Group to Harakat ul-Mujahedin.

Some chapters are less convincing than others. The one on Israel does not cover what might have been expected here – perceptions of Asian nuclear programmes and Israeli–Indian strategic cooperation. (It could also be argued that including chapters on Israel and Iran amounts to a very extensive and not entirely justified definition of 'Asia'.) The chapter on ASEAN countries states that 'Myanmar has not needed or sought nuclear weapons' (p. 457), an assertion that some would question. It says little about the extent of the US defence umbrella over Thailand and the Philippines. Also lacking is a chapter on proliferation networks and illicit nuclear trafficking on the continent, given what is known about imports and exports by Pakistan and North Korea, as well as the existence of a centrifuge-production factory in Malaysia for exports to Libya in 2001–03.

Alagappa has done much more than simply edit the book: he is also the author of four substantial analytic chapters. He discusses five findings: (1) nuclear weapons shape the political and strategic context in ways that go beyond their immediate security relevance (the 'long shadow'); (2) deterrence relationships in Asia are generally asymmetrical and not of a 'mutually assured destruction' type; (3) Asian nuclear states have widely differing goals and strategies; (4) these states have 'general deterrence' postures in the sense that they rely most of the time on implied and indirect threats; (5) extended deterrence is an important component of security on the continent.

The author correctly argues that there is little originality in Asian nuclear doctrines and postures, noting, for instance, that the 'tendency toward ambiguity and secrecy is not a peculiar Asian trait' (p. 25). He rightly calls for a common discourse and vocabulary on nuclear policies on the continent, which currently does not exist, a cause for misunderstandings in times of crisis.

Some of Alagappa's main conclusions are more problematic. He argues that nuclear weapons in Asia have 'reduced the salience of external balancing' (p. 26). The argument may work better the other way round: the absence of credible security guarantees played a significant part in decisions to go nuclear in Asia. And while it is certainly true that nuclear weapons have contributed to the strategic independence of countries like China and India – a revenge against colonisation – Pakistan's nuclear status has not reduced the importance of its relationship with China, and nuclear developments in North Korea and China have increased the importance of the US nuclear umbrella for Japan.

Alagappa argues that, overall, nuclear weapons have contributed to security and stability in Asia. One could counter that the author sees the glass as being half-full, but fails to see that it is also half-empty. It is true that there has not been a major military state-to-state conflict on the continent for a long time. But the combination of nationalism and nuclear programmes, in the context of rivalries for resources and territorial disputes, is a dangerous one. And the South Asian crises of 1999 and 2002 are reminders that nuclear powers can engage in dangerous behaviour when they feel protected by a nuclear umbrella. The absence of stable dyadic strategic relationships, the lack (at least for now) of assured second-strike capabilities in all countries concerned, and the multipolar nature of the Asian nuclear landscape, would seem to call for a much more measured assessment. To state that Asian nuclear countries 'have not behaved differently from the "rational" Western States' (p. 514) is cold comfort to anyone who has read extensively about the 1962 Cuban missile crisis.

Kalashnikov Culture: Small Arms Proliferation and Irregular Warfare
Christopher Carr. Westport, CT: Praeger Security International, 2008. £41.95/$75.00. 185 pp.

Merchant of Death: Money, Guns, Planes, and the Man Who Makes War Possible
Douglas Farah and Stephen Braun. Hoboken, NJ: John Wiley & Sons, 2007. £9.99/$15.95. 308 pp.

Small arms and light weapons kill thousands every day, but are much more difficult to control than strategic arms. The reasons are obvious: they exist in much greater numbers, are generally in the possession of individuals, and are increasingly produced by private or semi-private entities. The successful negotiations which led to the ban on landmines and cluster bombs could hardly be a model for their limitation.

As Christopher Carr's valuable study shows, the proliferation of small arms and lights weapons is also a multidimensional problem with important societal, cultural and economic dimensions. Carr uses the expression 'Kalashnikov culture' (of Pakistani origin) to describe areas of the world where these weapons not only fuel constant low-intensity conflict but are also at the centre of a social regime that flourishes when the state is dysfunctional and 'the reality of authority is invested in the lords of chaos, those who have mastered the climate of insecurity and violence and leveraged it into a peculiar form or opportunity' (p. x). Regions affected by Kalashnikov culture include large parts of Africa, Asia and the Middle East, where it was sometimes grafted onto traditional tribal cultures.

The appeal of small arms and light weapons is severalfold. They provide security but are also symbols of power, respect and manhood. They are often at the centre of a parallel economy that involves the culture of drugs, the predation of natural resources, and kidnapping. While they are not in themselves a cause of conflict, and can sometimes be just another tool that replaces other weapons for maintaining social order (as in Yemen or the Afghanistan–Pakistan region, for instance), they help raise crime and murder to previously unknown levels, especially in regions with 'youth bulges', where being part of a violent gang is often a way of life. Carr usefully distinguishes between Kalashnikov cultures and 'gun cultures' (as in Switzerland or the United States), where the large distribution of weapons poses little if any threat to the central government, their possession is regulated, and social norms prevent their large-scale use.

While the responsibility for the proliferation of small arms and light weapons cannot be attributed to any specific actor, some individuals and countries have played a major role in this phenomenon. Viktor Bout, the über-arms dealer made famous by the movie *Lord of War* (2005), is one of them. A Russian citizen whose base of operations was the small emirate of Sharjah, Bout is a poster child for the illegal arms trade of the 1990s, during which he exported enormous quantities of weapon systems, large and small, as well as tonnes of ammunition, from Eastern Europe to Africa. He perfectly matched the offer – the vast arsenals of the former Warsaw Pact – with the demand, created by conflicts spawned by the combination of abandonment by great powers after the Cold War, abysmal state failures, competition for natural resources and youth bulges. A key to his success was building the largest private fleet of ex-Soviet cargo aircraft in the world, which he registered in countries such as Liberia to escape international attention.

Merchant of Death tells the story of the hunt for Bout by intelligence and law-enforcement agencies, and the ability of their target to outpace the often badly coordinated multinational effort. Sadly, this was made all the easier by the fact that many global players actually employed his network in one way or another, including Western and even UN contractors, who used his cargo planes. At the turn of the century, two individuals exemplified the dark side of globalisation and new forms of transnational threats: Osama bin Laden (violent jihadism) and Abdul Qadeer Khan (nuclear proliferation). A third name deserves to be added to this list, that of Viktor Bout – arrested in Thailand on 6 March 2008 – for his contribution to human suffering and misery.

The Great American Gamble: Deterrence Theory and Practice from the Cold War to the Twenty-First Century
Keith B. Payne. Fairfax, VA: National Institute Press, 2008.
$19.00. 471 pp.

This is a twenty-first-century rarity: a nearly 500-page treatise on nuclear strategy. While such a description might put some readers off, the book is worth a look. Although about half of this heavy volume is devoted to a methodical deconstruction of the United States' Cold War strategic-deterrence paradigm, it is resolutely forward-looking, as author Keith Payne's ambition is to ensure that deterrence remains relevant and efficient. And as one of the very few experts on the subject who has been influential both outside and inside government – he was one of the main authors of the US 2001 Nuclear Posture Review – Payne is a trusted analyst who deserves to be heard.

Payne's study builds upon his earlier work, and occasionally uses material from previous books and articles. The author invites us to re-read the evolution of nuclear strategy by using a simple and effective prism. According to him, there were always two competing schools of thought in the Western strategic debate, one exemplified by Thomas Schelling, the other by Herman Kahn. The first argues that deterrence is best achieved by a 'stable balance of terror', based on an assured destruction capability on both sides (and thus the absence of significant defences), which minimises the risk of surprise attack. Within such a framework, the risk of aggression is discouraged by the mere possibility that escalation could easily get out of control (the 'threat that leaves something to chance'). Such fear makes extended deterrence possible.

The second school argues that deterrence requires an 'imbalance of terror', where nuclear superiority and strategic defences make the defender able to deny any advantage to the aggressor, in particular by ensuring an ability to ensure 'damage limitation'. According to this analysis, deterrence – in particular extended deterrence – can only be ensured by inducing a rational fear of escalation in the enemy's mind, a 'threat that leaves little to chance', as Payne, who clearly states his own preference for this form of deterrence, puts it (p. 38).

As is well known, the first school overwhelmingly won the day and became the 'dominant norm' in the early 1960s. Payne convincingly shows that it has never left this position, and that later attempts to adapt US nuclear strategy were more 'variations' than real alterations.

It is somewhat difficult to side with Kahn as decisively as Payne does, because one of Schelling's crucial contributions was to integrate the 'fog of war' in the nascent theory of nuclear strategy, which otherwise might have run the risk of being an exercise in intellectual elegance and mathematical certainties irrelevant to the real world. And Payne himself gracefully acknowledges many strong points in Schelling's theories, arguing that his later followers were 'dumbing down' his complex thinking (p. 252). But preferences for one school of thought or the other rely on different expectations of the behaviour of adversaries in a nuclear war, something that has never been tested and will never be if deterrence succeeds.

Payne is rather more convincing as he seeks to shape the contours of a deterrence strategy for the coming decades. He warns against the simplicity of the 'stable deterrence' model, which can only be valid in very specific conditions. He gives countless historical examples of the failures of deterrence and of seemingly irrational (or hardly predictable) behaviour by governments. He warns against generalisations and postulates in deterrence thinking, arguing in favour of 'tailored' deterrence relying on intimate knowledge of the adversary. He is on the mark when he states that '"How much do you know?" must precede "How much is enough?"' (p. 303). He makes a strong case for missile defence, arguing that it may be a small price to pay as an insurance against the failures of deterrence. This will have to be judged by the taxpayers of the United States and its allies. But as the global nuclear picture becomes more complex and Western analysts wonder whether and how China, North Korea and Iran could be deterred, Payne's volume is an invaluable guide for methodical thinking on the subject.

Asia-Pacific
Lanxin Xiang

China's New Confucianism: Politics and Everyday Life in a Changing Society
Daniel A. Bell. Princeton, NJ: Princeton University Press, 2008.£15.95/$26.95. 240 pp.

'A joke about China is that one can say anything about it without getting it right. Another joke about China is that one can say anything about it without getting it wrong.' With these sentences, Daniel A. Bell, a Canadian scholar teaching in Beijing, begins *China's New Confucianism: Politics and Everyday Life in a Changing Society*, the result of years of observing China at close range.

Does Bell himself get it right or wrong? As a good China scholar, he is too self-conscious to adopt the typical Western habit of depicting China as black or white, as do many books of this kind that are hitting bookshelves too rapidly these days. His perspective is promising: he looks at the real debates among Chinese rather than those among Westerners. The latter – about such issues as Tibet, human rights and democratisation – often attract media attention, but are irrelevant to Chinese realities. Instead, Bell focuses on what he calls, albeit somewhat awkwardly, the 'new Confucianism' in Chinese politics and daily life, drawing attention to Chinese intellectual debates on some fundamental political and social issues.

The book is divided into three parts. In the first, Bell describes the changing discourse on China's political future. Gone are the days when 'democracy' was the only focus of political debate among the Chinese intellectual elite. A new question has emerged: does Confucianism pose a serious challenge to Western-style liberal democracy? 'There are reasons', says Bell, 'to think that they are compatible, if not mutually reinforcing.' Even Wang Juntao, a liberal democracy activist and former political prisoner in China, accepts this argument. What is even more interesting is the growing debate over foreign affairs. As public debates on foreign policy – a strict taboo in Mao's day – have become more open, the Chinese elite have found a new outlet to express themselves and have rediscovered such useful Confucian concepts as *tianxia* (the cosmopolitan ideal) and *hexie* (harmony). Chinese leaders frequently use these terms as part of their intense campaign to promote China's newly acquired soft power.

Part Two discusses China's changing society. Here, Bell raises a number of important issues. For example, instead of labelling prostitution as the dark side of the economic boom, Bell argues that the Chinese sex industry has promoted a revival of the traditional concubinage system, especially in the strange form of the 'singing and sex' trade, a variation of the widespread karaoke business. The net effect of this system is not always negative in terms of economic benefits and social stability.

Part Three is about education. Contrary to the common belief in the West, Bell finds in China a paradise of academic freedom. He pointedly compares the academic atmospheres of China and Singapore, where he previously taught and apparently had a bad experience, noting that teaching Western political theories incurs fewer political problems in China than in Singapore.

Bell is particularly good at dissecting several key Western misperceptions about the 'Confucians' in China, who are often thought to be 'old, serious and conservative'. Confucianism is above all a way of life, and the younger generation is not averse to it any more. In Bell's view, a Confucian need not be politically conservative, but must be conscious of both family and community obligations. Thus, Bell does not see much hard evidence that social and political instability will ensue when and if the Chinese body politic revives Confucianism.

China's New Confucianism stands out for not conforming to a preordained Western conceptual framework. The personal anecdotes are interesting and Bell displays cultural sensitivity throughout, including in his many references to his Chinese wife. But though the book is a worthwhile read for anyone interested in a rapidly changing China, its argument is already being eclipsed by events. Current Chinese debates have turned in a new and more important direction: they are now over far more profound issues than the new Confucianism, including 'universal value', economic justice and the international system. These debates seem more in line with the Cambridge school of conceptual history than the old and rather prosaic debates over freedom and democracy. In other words,

Chinese debates are moving rapidly into a postmodern phase, while the rest of the world still looks at China from the modern perspective of the European Enlightenment.

The Japanese Challenge to the American Neoliberal World Order
Yong Wook Lee. Stanford, CA: Stanford University Press, 2008. £43.50/$60.00. 284 pp.

It is interesting that this book, written by a scholar who teaches at the University of Oklahoma, reverses the debate launched in the 1980s, when Japan-bashers in the United States were attacking Japan's unwillingness to embrace liberal economic principles at home and abroad. Yoon Wook Lee thinks that the Japanese actually got their revenge after the Cold War, when Anglo-American triumphalism fed Western calls for a neoliberal world order. By 'bringing the state back in' to economic development in the mid 1980s, Japan made a deliberate effort to challenge the neoliberal orthodoxy. This is the 'Japanese challenge' of the title, which the author believes was 'aimed at undermining U.S.-led neoliberal attempts to delegitimate the role of the state in economic development'.

Efforts to promote a state-led alternative model of economic development only intensified in the 1990s. Lee's discussion of Japan's proposal to establish an Asian Monetary Fund (AMF) during the 1997 Asian financial crisis, for example, is especially interesting. After initially cooperating with the International Monetary Fund (IMF) to address the Thai crisis in May 1997, Japan abruptly made the proposal for an 'Asian only' monetary institution to end IMF interference.

Japanese economic theory stresses that state action in the economic sphere is normal in the context of global economic history. By contrast, the laissez-faire ideal which has sustained the Anglo-American faith in the 'magic of the market' is actually ahistorical, abnormal and idiosyncratic. Among the three dominant ideologies in postwar Japan, Marxism, economic liberalism and developmentalism, it was the latter that ultimately triumphed, giving rise to the Japanese challenge. In particular, the Japanese developmentalists accuse neoliberals of hijacking the history of the world economy.

Indeed, the roots of Japanese challenge can be traced quite far back in the country's history. After all, the Japanese state – in accordance with a pattern commonly seen in late industrialisers – has been active in the country's economic affairs since the Meiji Restoration. Hence, rather than seeing 'universal value' in market forces, perhaps it is state-assisted development that should be considered the universal value. The author concludes that this is part of the shared 'Asian identity', an analysis based on political scientist Alexander Wendt's 'identity–intention' framework. Thus, Lee provides a new perspective not only on US–Japan relations, but also on the whole phenomenon of Asian regionalism.

East Asian Multilateralism: Prospects for Regional Stability
Kent E. Calder and Francis Fukuyama, eds. Baltimore, MD: Johns Hopkins University Press, 2008. $25.00. 281 pp.

This slim volume is a timely contribution to the current debates in Asia on a rapidly emerging regionalism. The editors and contributors are an interesting mixture of distinguished scholars, graduate students and policy practitioners. In the book's introduction, the editors point out that a rising East Asia displays an 'organization gap', or an absence of formal and multilateral organisations to reflect regional trends towards economic interdependence.

According to Kent Calder, the origins of this gap can be traced to the Korean War. Two regionalist schemes emerged after the Communist victory in the Chinese Civil War, the Pacific Pact and the San Francisco System. The former, based on a proposal by Philippine Foreign Minister Carlos Romulo in January 1949, was originally conceived as a league of Asian nations addressing issues of economic development as well as security. But the project took a military turn in 1950 when the idea of a Pacific Pact emerged as a multilateral military alliance, or a mini NATO in Asia-Pacific.

An alternative plan for a regional security system, first proposed at around the same time, eventually developed into a new, second-best system after the outbreak of the Korean War. This so-called San Francisco System abandoned multilateralism; instead, it was based on a set of bilateral security arrangements or understandings between one power, the United States, and other individual players in the region. The most important feature of this system, compared with the Pacific Pact, was a commitment to gradual economic integration of the allied economies on preferential terms; hence an explicit economic dimension was established. It was this system that eventually took hold in the region.

Both the Pacific Pact and the San Francisco System were direct reactions to the so-called 'Loss of China' to the Communists. Today, the lack of a multilateral organisation in the region is becoming an urgent problem, and once again China is at the heart of regional concerns, this time because of its participation in the world system rather than its absence. More importantly, after decades of practicing unilateral policies, China itself has moved toward multilateralism. The 1997 Asian financial crisis and the Six-Party Talks over North Korea's nuclear-weapons programme are two instances where Chinese participation has been of critical importance. So the question is whether East Asians might finally close the organisational gap and establish a formal economic and security institution in the region.

Kent Calder, who provides a useful historical survey of Northeast Asian regionalism, seems rather sceptical of the prospect of a multilateral future, pointing to what he calls a 'trust gap' among the chief players and the continued presence of the main features of the original San Francisco System. Bruce Cumings is equally sceptical of a smooth transition to multilateralism given that the region has so far been dominated by unilateralism. The United States remains the key enabler of either multilateralism or unilateralism in the region, and unfortunately, the Bush administration would prefer to apply its seemingly effective unilateralism in East Asia to the rest of the world, rather than foster Asian multilateralism.

The chapter on China by Cheng-Chwee Kuik is interesting, for it makes a strong case for an 'innenpolitik' (domestic-policy) interpretation of China's motivation for multilateralism. Since the prime concern of the Communist regime is the preservation of its domestic dominance rather than the maximisation of its external power, China is not obsessed with the idea of replacing the United States as a new hegemonic power in the world. This undercuts the China Threat thesis, popular in many quarters of US society for years. But Kuik does not write from the perspective of a Chinese insider, so his chapter should not be seen as providing a Chinese angle on the question.

Francis Fukuyama concludes that the United States may face major strategic choices between bilateral or multilateral frameworks, pro- or anti-Chinese postures, and US inclusion in or exclusion from regional affairs. But he believes that these choices are by no means as stark as they appear, and it is possible for the United States to combine some of these choices to make the Asian system work. Although this collection of essays neglects an explicit perspective from the Chinese side, it is a worthwhile read for anyone interested in recent momentum towards regionalism in East Asia.

US Taiwan Policy: Constructing the Triangle
Oystein Tunsjo. Abingdon and New York: Routledge, 2008.
£75.00/$145.00. 194 pp.

Oystein Tunsjo, a senior research fellow at the Norwegian Institute for Defence Studies, claims his book offers 'the first rigorous and detailed critical constructivist analysis based on original and detailed archival research of the construction of the Taiwan issue in US China policy'. His purpose is to challenge mainstream approaches, especially as embodied in the standard works by Harry Harding (*A*

Fragile Relationship), Rosemary Foot (*The Practice of Power*), David Shambaugh and Thomas Robinson (*Chinese Foreign Policy*), and others.

Tunsjo succeeds to a large extent. He points out that most books offering 'realist analysis' are merely based on international-relations theories of rational choice, but not on sufficient archival evidence. The kind of interest-based, balance-of-power interpretations offered in mainstream literature is not just outdated, but often misleading, as if national interests were objective, hard realities that could be studied through a method similar to that of natural science. According to the author, the traditional school asks 'why' questions, while the constructivist approach asks only 'how-possible' questions.

From a constructivist perspective, the origins of the Taiwan issue must be traced to the decision by the Truman administration to intervene in the conflict in the Taiwan Strait between a communist China and the nationalist regime on the island of Taiwan in 1950, immediately after the Korean War broke out. Instead of attributing this decision vaguely, and without much archival evidence, to the alleged necessity of protecting US geopolitical interests, as many authors have done, Tunsjo raises the question of how it was possible for the Truman administration to reverse its policy of abandoning the Chiang regime by instituting, on 27 June 1950, a 'neutralising' strategy in the Taiwan Strait. How did a North Korean attack on South Korea translate into this kind of major policy reversal? The decision, according to the author, was not so much a realpolitik calculation as a symbolic move. The Korean War changed US perceptions of the Chinese Communist regime, from a potential Titoist rebel within the communist bloc to a leading 'red menace'. The status of Taiwan became a symbolic issue in the quest to stop communist expansion across the entire Asian region.

Above all, Tunsjo emphasises the contingent nature of US Taiwan policy. This contrasts with many standard works, which are based on assumptions of pre-determined national interest. Tunsjo's approach not only widens our understanding of decision-makers' scope of maneuvering, but also reminds us that shifts in US Taiwan policy in the past, as in the future, may not be explained by fixed factors or rational choice at all. The future of Taiwan may well point in a direction that previous generations would have considered unthinkable.

This is a small book, but the author has made an important contribution to the study of US Taiwan policy and the relationship between China and the United States in general. It 'de-centres', to use Jacques Derrida's favourite expression, the entire conceptual framework on this issue, and debunks the various rational-choice approaches, too often demonstrated by leading American diplomatic historians, who tend to assume a clear interest-based US policy toward the

Taiwan question without taking into consideration the various discourses that have surrounded it, both in United States and abroad. By emphasising historical contingency, Tunsjo elevates diplomatic history to a far more sophisticated level, allowing international-relations experts to converse with contextualists such as Ludwig Wittgenstein or Quentin Skinner.

Taiwan's Relations with Mainland China: A Tail Wagging Two Dogs

Su Chi. Abingdon: Routledge, 2008. £75.00/$150.00. 342 pp.

Su Chi, a prominent scholar and member of Taiwan's ruling party, the Kuomintang, has written an intriguing book on the Taiwan question. From his vantage point as a policy practitioner and sharp observer, Su offers not only a first-hand account of Taiwan's relationship with mainland China, but also a sensible analysis of one of the most complicated relationships of our times. Su is currently secretary-general of Taiwan's National Security Council and served as cabinet-level chairman of the Mainland Affairs Council in the 1990s. This book grew out of a Chinese-language book, *Brinkmanship: From Two-States Theory to One Country on Each Side*, published in Taiwan in 2003. He has added new materials and revised the book substantially for publication in English.

The story begins with a fascinating account of secret diplomatic exchanges between the two sides, many details of which were unknown to the outside world until now. These secret overtures in the wake of the end of the Chiang dynasty achieved remarkable success. They resulted in major breakthroughs, particularly the Koo–Wang talks in 1993. This breakthrough was predicated on the principle of 'one China, respective interpretations'. After 1993, however, the story took a dramatic turn when Lee succeeded in obtaining a visa from the US government to visit his alma mater, Cornell University. The previous spirit of rapprochement was replaced by confrontation, and mainland China launched fierce campaigns against Lee, even making military threats in the form of missile exercises in 1995–96.

The United States had the most active reaction to this campaign, moving to reduce tension in the Taiwan Strait by sending aircraft-carrier groups to the region. The Taiwanese regime reacted only passively. But at the end of the decade, Lee regained the policy initiative by announcing a 'Two-State Theory' in 1999, which further angered mainland leaders. The situation deteriorated even more dramatically when the pro-independence leader Chen Shui-bian ascended to power in 2000. Chen clearly wanted *de jure* independence for Taiwan, a stance that forced the two big players, Beijing and Washington, into yet another round of frantic diplomatic manoeuvres, a situation neither side wanted. Thus, according to the author, during the period from 1988 to 2004, Taiwan was a tail that could wag two big dogs.

Beijing's resort to the use of force in response to Lee's Cornell visit was clearly a mistake driven by desperation. But Chen also behaved provocatively, especially after a pro-Taiwan Bush administration came to power in 2001. Chen made an extremely bold move beyond the 'Two-State Theory' in 2003, announcing a 'one country on each side' policy, together with a set of tactical steps including 'referendum', '(state) name change', and 'writing of a new constitution'. When he won a second term after surviving a dubious assassination attempt, Chen thought he had obtained legitimacy and a popular mandate for making the final push towards independence. But political intervention from the United States and military actions from the mainland jointly foiled Chen's scheme and Taiwan's status was thus maintained by the two big dogs.

The author identifies three impulses that drove these events: the use of force, unification and independence. Combined they formed a combustible mix, which could have resulted in an explosive confrontation. But Su believes they produced a vicious cycle instead, in which none of the impulses could dominate. Thus, Beijing failed to achieve unification through the use of force, and Taiwanese leaders failed to move in any substantial way toward independence. The United States, on the other hand, appears more successful as a skilful balance-holder in this three-way game. Washington's macro- as well as micro-management of Taiwan affairs seems to have been surprisingly effective. It is an intriguing analysis, and of interest to anyone concerned with cross-Strait relations.

Closing Argument

African American

Jonathan Stevenson

I

During the decade between the end of the Cold War and the 11 September attacks, America's Africa policy was more about political and ethical values than strategic interests. In 1992, a lame-duck President George H.W. Bush bet that humanitarian intervention in Somalia would help to establish a kinder, gentler 'new world order'. It didn't pay off, and made the Clinton administration gun-shy about intervening in Rwanda, where timely major-power action might have prevented the genocide. During the latter part of the 1990s, the United States handled instability in West African states mainly with measured participation in an international campaign against illicit 'blood diamonds'. By denying destabilising elements the source of their lucre, this effort facilitated UN-managed political reconstruction and relative peace in Sierra Leone and Liberia, but hardly constituted full-fledged engagement. Washington's response to the HIV/AIDS pandemic was (and remains) founded in humanitarian concern, not strategic alarm. Although state failure in Somalia and the Democratic Republic of the Congo (DRC) and Robert Mugabe's malign and incompetent governance in Zimbabwe drew rhetorical consternation, Washington largely deferred to weak multilateral organisations and feckless regional powers respectively on these matters. Sustained US bilateral engagement was generally limited to 'anchor states' such as Nigeria, South Africa and Kenya.

Jonathan Stevenson, a Contributing Editor to *Survival* and Professor of Strategic Studies at the US Naval War College, was deployed on the USS *Fort McHenry*, lead element of the Africa Partnership Station, in April 2008.

Survival | vol. 51 no. 1 | February–March 2009 | pp. 249–256 DOI 10 1080/00396330902749848

During George W. Bush's tenure, concerns about Middle Eastern stability yielded substantial US counter-terrorism assistance to North African governments and Washington's fruitful if limited diplomatic engagement of Libya. Bush also substantially increased direct humanitarian and development aid to Africa. But Washington's preoccupation with Iraq has meant diminished strategic attention to sub-Saharan Africa. The management of US interests in West Africa – which consists mainly of maintaining access to hydrocarbons – has essentially devolved to oil multinationals. Washington has for the most part limited US involvement in East Africa to counter-terrorism, including support for a politically short-sighted take-down of Somalia's de facto Islamist government by Ethiopian forces; the episode raised anti-Western sentiment among Somalis and conditioned an Islamist resurgence.[1] Owing in large part to the American Christian Right's support for the Sudanese Christian rebels in southern Sudan, the White House has advanced resolution of the conflict there. But the January 2005 peace agreement between Khartoum and the southern Sudanese rebels, which the United States helped broker, is in danger of unravelling. Not even a State Department finding of genocide in Darfur has been enough to prompt the serious commitment of an overstretched US military to a UN–African Union peacekeeping effort.

II

A mandate for a more humble and consensual foreign policy could dovetail with the first African-American president's political and personal inclination to reinvigorate Africa policy. To some considerable extent, the Bush administration has already recognised Africa's strategic importance, establishing an entire new military component – Africa Command (AFRICOM) – dedicated to the continent.[2] Moreover, the disturbances that have beset Kenya, Sudan, Zimbabwe, Somalia and the Maghreb in 2008, and the potentially corrosive effect on European security of rising jihadism in the latter two areas, make the need for attention from major powers objectively clear. Among the former colonial powers, only France – with a diplomatic and military intervention in Côte d'Ivoire and a major troop contribution to the UN peacekeeping force in the DRC – has made Africa a priority. Ideally, European governments will join the United States in a new commitment to Africa.

Increasing numbers of aspiring jihadists are choosing Somalia and Algeria – as well as Pakistan – as training grounds instead of Iraq, where they calculate that the odds of surviving to fight another day are prohibitively low.[3] Weak and failed states are vulnerable to co-optation by bad actors, and there are more of such states in sub-Saharan Africa than anywhere else. The two of greatest concern are Islamist-governed Sudan and anarchic and increasingly Islamist Somalia, from which pirates, unchecked by effective local or national authority, have been wreaking increasing havoc on commercial shipping in the Gulf of Aden and Indian Ocean. Sub-Saharan oil and gas reserves, though insufficient to end American dependence on Middle East supplies, are substantial enough to warrant durable attention to potential causes of their insecurity or inaccessibility, including competing customers.[4] In particular, China's rapidly growing economy is increasingly thirsty for foreign oil; 28% of its oil imports come from Africa. By virtue of its interest in energy security, trade and investment opportunities, and diplomatic legitimacy, Beijing has put Africa back into geopolitical play. Meanwhile, African governments have been striving to ameliorate the continent's chronic poverty, disease, bad governance and state incapacity, and asking the United States and Europe to meet them halfway via aid-channelling mechanisms such as the New Partnership for Africa's Development (NEPAD) led by South Africa and Nigeria.

The new American administration, then, is unlikely to be able or inclined to sidestep Africa. American values and interests are converging there, but US Africa policy must balance the two more effectively. The Bush administration's signature attempt to harmonise interests and values in foreign policy – the invasion and occupation of Iraq – has been a failure mainly because the interest (counter-proliferation) was illusory and the value (democracy promotion) unavailing. One consequence has been the palpable desire for returning to a more realist foreign policy tailored to vital interests and sceptical of grand ideological solutions. Another has been the recognition that America's unipolar moment has passed – or been squandered – and that US policy therefore must be implemented with a lighter touch. Yet, at the same time, it is infeasible – ethically, if not diplomatically – for the United States to drastically compromise its existing governance criteria for engaging African regimes. The challenge is for the United States to hold the ethical high ground while creating incentives for African states, and perhaps even geopolitical rivals, to do the same. But China is less concerned with these values than the United States. By offering aid with no strings attached other than the recipient's alignment with China's primarily economic interests – for instance, by building oil refineries in Sudan to encourage Khartoum to enter into a preferential supply relationship with China, and providing military and economic assistance to Mugabe's government in exchange for mining concessions in Zimbabwe – Beijing has allowed abusive regimes greater leeway. At the same time, China's attainment of its long-range objectives in Africa – securing access to natural resources, expanding export markets, burnishing its credentials as an international political player, and preserving its own international freedom of action – ultimately depends on a stable and secure regional environment, which imposes limits on its more objectionable practices.

The United States and its allies can counter Chinese influence by offering African governments needed services and products that only they can supply and continuing to link them to broad standards of good governance. Along these lines, the US Navy and the former European colonial powers it has enlisted as partners, including France, the United Kingdom, Portugal and Germany, have already begun to project power in a non-confrontational way to improve maritime security in the Gulf of Guinea. According to Navy thinking, as long as outright armed intervention isn't necessary, it makes more sense to reach out to foreign governments and populations itinerantly, from ships, without burdening them politically and intimidating them psychologically with a big American ground presence. The Africa Partnership Station (APS) is a group of US Navy ships staffed with European and African as well as American officers that undertakes long-term patrols with frequent but short-term stops. The intent is to establish relationships with foreign officials that allow the United States to exert positive influence while avoiding the tension that an in-country US presence might produce. Operational training courses for West African navies and coast guards aim to build the strongest and most durable ties, and community outreach programmes (repairing orphanages and building schoolhouses and medical facilities; delivering meals to feeding centres and medical supplies to clinics and hospitals; visiting schools, churches, and mosques) to improve the diplomatic milieu and warm local attitudes.

The APS's inaugural six-month tour in 2007–08 also highlighted structural inadequacies in US Africa policy: while the Navy showed its skill as a policy entrepreneur, coordination among other agencies lagged. For example, US diplomacy and strategic communications proved insufficient to win the complete buy-in of Nigeria, a country whose participation in the APS is essential to its overall success. The APS, of course, constitutes only a small-bore model for any comprehensive strategic outreach. But the inter-agency lethargy that affected the APS's first iteration seemed to mirror more general incoherence. The US Agency for International Development (USAID), for instance, has been hollowed out and disempowered during Bush's presidency, with the Pentagon administering most foreign-assistance programmes and the State Department running the Millennium Challenge Corporation and the President's Emergency Plan for AIDS Relief. Declining revenues due to the economic crisis will slow US foreign assistance to the developing world. But by reviving USAID and elevating it to cabinet rank – along the lines of the UK's Department for International Development – Obama would signal the developing world, and Africa in particular, that the United States intends to give its welfare a more central place in US policy.[5]

Washington should be clear about its wider strategic intentions and constraints. In particular, it will have to be candid about the reality that certain US ground-based military efforts in Africa may make sense on the basis of US interests, values, or both. In continuing to clarify the uses of AFRICOM for public consumption, Washington should acknowledge openly and clearly that two of the new command's biggest challenges may end up being sustaining energy security for mutual benefit, and peace-keeping and state-building, which the Pentagon is also weaving more thoroughly into US military doctrine.[6] Washington should note further that AFRICOM will provide the United States with the bureaucratic means for enhancing diplomatic and military-to-military relationships with key

African states and African regional organisations so as to better meet these challenges. To discourage perceptions of prospective US domination, the US government should also stress its official preference that African forces or UN peacekeeping contingents be used in African territory. Yet it should harbour no illusions about Africa's indigenous peace-enforcement and peacekeeping capacity, which will be inadequate for meeting the continent's needs for some time to come.

For most if not all of Obama's first term, however, the lion's share of US military assets will be needed in Iraq and Afghanistan. (Most of NATO's assets will be needed in Afghanistan as well.) Large-scale American and European military contingents will be unavailable for Africa. Major powers can compensate with enterprising diplomatic initiatives for resolving conflict and alleviating longstanding tensions, both within the region and between major and regional powers. Washington in particular could pursue a more vigorous policy of engagement with a wider range of African governments, including that of Sudan, whose military weakness and internal political fragility may make it more receptive to American inducements.[7] Even Somalia should not be ruled out as the object of US conflict-resolution efforts.[8] Indeed, if Islamists continue to regain territory and political traction and keep targeting the relatively peaceful and successful breakaway state of Somaliland in the northern part of the country, and Somali piracy persists in imperiling strategically vital commercial vessels like oil tankers and freighters carrying food, a diplomatic initiative may become an urgent priority.[9] Whether such efforts can work will depend on the degree of multilateral backing that can be mustered and, to the extent that the United States is involved, how much credibility Washington has with Africans.

III

US diplomatic leverage in the region will depend critically on whether African governments perceive the United States' use of military power in Africa as judicious or heavy-handed. On this score, one of AFRICOM's most potent tactical assets, the special-operations forces of the 2,000-strong Combined Joint Task Force–Horn of Africa (CJTF-HOA), based in Djibouti, could become a strategic liability if the United States is not cautious about its employment. CJTF-HOA has collected intelligence and bolstered regional governments' internal-security capabilities and civic programmes. But CJTF-HOA also helped train and supported the Ethiopian troops that, with Washington's tacit approval, in late 2006 and early 2007 overthrew the Islamic Courts Union, an Islamist coalition with links to al-Qaeda that had taken de facto control of southern Somalia six months earlier. Subsequently, several US attacks on suspected terrorist targets in Somalia with AC-130 gunships claimed civilian casualties, to the consternation of many Africans fearing a neo-colonial and militaristic US Africa policy.

So the suggestion of some American officials that AFRICOM's future mode of engagement should be modelled on CJTF-HOA's local capacity-building mission seems misguided.[10] Historically and functionally, CJTF-HOA is mainly a hard counter-terrorism tool, and thus unavoidably embodies an emphasis on the use of force.[11] The same goes for Navy ships positioned in the Indian Ocean mainly to monitor terrorists and target them when the opportunity arises, as it did in May 2008 when a cruise-missile strike targetted terrorist suspects in central Somalia. Assets like CJTF-HOA would be best used not as 'kinetic' counter-terrorism tools but as mechanisms for ramping up train-and-equip programmes for African security forces, bolstering foreign internal defence, and perhaps fortifying humanitarian relief and peacekeeping efforts.[12]

The advent of AFRICOM has raised fears among Africans that the United States will adopt a neo-colonial Africa policy and 'militarise' its relationship with the continent in a way that caters mainly to short-term American security and economic needs and crowds out long-term African ones.[13] By using AFRICOM with purposeful restraint, and in coordination with pure soft-power instruments like USAID, the new American president can calm African worries by keeping the United States' military footprint relatively small; emphasising hands-on help in the form of foreign assistance, security training, conflict resolution and peacekeeping; and conducting America's African business to the greatest extent possible multilaterally, through African regional organisations and the United Nations. Implementing the national policy of energy independence that Obama set forth during his campaign will also lend credibility to US claims that it is not in Africa merely for the oil. Obama, as the son of a Kenyan and the first African-American president, will enjoy extraordinary presumptive credibility and good will among African populations and governments. To maintain it, he will have to conduct US policy not merely to advance America's immediate interests but also to better integrate the strategic futures of Africa and the West.

Notes

1 See Ken Menkhaus and Karin von Hippel, 'Republic of Blowback', *International Herald Tribune*, 4 September 2008; Jonathan Stevenson, 'The Somali Model?', *The National Interest*, no. 90, July/August 2007.

2 See, for example, 'AFRICOM: The United States' New Combatant Command', IISS, *Strategic Comments*, vol. 13, no. 2, March 2007.

3 See, for example, Michael Evans, 'Al Qaeda Finds Three Safe Havens for Terror Training', *Times*, 2 July 2008.

4 See Jonathan Stevenson, 'Africa's Growing Strategic Resonance', *Survival*, vol. 45, no. 4, Winter 2003–04, pp. 153–72.

5 See J. Brian Atwood, M. Peter McPherson and Andrew Natsios, 'Arrested Development', *Foreign Affairs*, vol. 87, no. 6, November/December 2008, pp. 123–32.

6 See Mark Bellamy, Kathleen Hicks and J. Stephen Morrison, 'Strengthening AFRICOM's Case', Center for Strategic and International Studies, 8 March 2008; http://www.csis.org/media/csis/pubs/080305_africom.pdf.

7 See Andrew S. Natsios, 'Beyond Darfur', *Foreign Affairs*, vol. 87, no. 3, May/June 2008, pp. 77–93.

8 See Jonathan Stevenson, 'Risks and Opportunities in Somalia', *Survival*, vol. 49, no. 2, Summer 2007, pp. 5–20.

9 See Mohammed Ibrahim and Jeffrey Gettleman, '5 Suicide Bomb Attacks Hit Somalia', *New York Times*, 29 October 2008; Jonathan Stevenson, 'To Do: Somalia', *The New Republic Online*, 21 November 2008; Jonathan Stevenson, 'Somali Pirates at Sea', Forbes.com, 19 November 2008.

10 See Thomas Barnett, 'Continental Ambitions', *Guardian*, 20 December 2007; Ginny Hill, 'Military Focuses on Development in Africa', *Christian Science Monitor*, 22 June 2007.

11 Stevenson, 'The Somali Model?'.

12 See Lauren Ploch, *Africa Command: U.S. Strategic Interests and the Role of the U.S. Military in Africa*, Congressional Research Service (CRS) Report for Congress, Order Code RL 24003, updated 10 March 2008.

13 See Darren Taylor, 'Skepticism Abounds as AFRICOM Launch Looms', *Voice of America*, 26 May 2008, http://www.voanews.com/english/archive/2008-05/Skepticism-Abounds-as-AFRICOM-Launch-Looms-PART-2-of-5.cfm.